D1338305

IDENTIFICATION OF THE LARGER FUNGI

DEDICATION

To my parents who encouraged my interests in mushrooms and toadstools and my wife who, later, was sympathetic to my studies and assisted in the production of the manuscript.

Hulton Group Keys

IDENTIFICATION OF THE LARGER FUNGI

by
ROY WATLING, B.Sc., Ph.D., M.I.Biol.
Principal Scientific Officer,
Royal Botanic Garden, Edinburgh

Editor of series: **Antony R. Kenney, M.A., B.Sc.**

©
1973
R. Watling
A. R. Kenney
ISBN 0 7175 0595 2

First published 1973 by Hulton Educational Publications Ltd.,
Raans Road, Amersham, Bucks.
Reproduced and printed by photolithography and bound in
Great Britain at The Pitman Press, Bath

PREFACE

This is one of a series of books intended to introduce field-biology to students, particularly the sixth form and early university student. The present work is ecologically biased in order to emphasise a rather neglected aspect of the higher fungi.

Few books on fungi have ever been designed for students. This book is aimed primarily at this level, but if the interested amateur is assisted and encouraged by this same text my hopes will have been doubly achieved. Many amateurs interested in higher fungi wish only to name their collections, or know approximately what they are before sampling them as an addition to their diet. An understanding of our commoner species at an early age will allow the 'budding' mycologist to tackle the much needed study of the more critical forms. Mycology is still at a descriptive stage, but it is hoped this will soon be changed and fungi of all kinds will be studied as part and parcel of courses in ecology.

It is of course quite impossible to cover all the species in such a small volume as this present one, but it is hoped that the examples which have been carefully chosen are sufficiently common throughout the country for any student to collect them in a single season. The examples, except for very few, in fact appear in the list of higher fungi found about the Kindrogan Field Centre, Perthshire, Scotland, compiled from the collections made by students attending my field course there.

The present work is arranged in three parts: the agarics are dealt with first, the non-agarics next, both with particular reference to their major habitat preferences, and lastly a catalogue of those more specialised habitats which are frequently encountered. All parts are supported at the end by lists in tabular form of those species expected to be found in any one habitat. Keys to the major groups, families and genera, are included to widen the scope of the book and place the examples chosen and illustrated in the text in their position in classification.

In the description the synonymy has been very severely pruned and only covers the commonly seen names; they are included as part of the general information under each species. In order for the student to expand unfamiliar names a list of references is added at the end of the work. The common names of the fungi, whenever possible, have been adopted from a list produced by Dr Large, the author of *The Advance of the Fungi*, an exciting tale of fungal parasites. The authorities for the names of the fungi described have been reduced to accord with the minimum requirements set out by the Code of Botanical Nomenclature. After each description a list of references to coloured plates is given and while some of these illustrations are not of the highest quality they are adequate, and, more important, they are widely available. Any technical terms appearing in the description are explained in the glossary, although they have been kept to a minimum; the difficulty of expressing colours has been overcome by consistently referring to one colour chart only, (a chart designed originally for the use of mycologists and available from Her Majesty's Stationery Office).

I have not indicated the edibility of a particular species unless there is no doubt as to the edibility of it, related species and those species with which it might be easily confused. Many fungi are notoriously difficult to identify and when one has approximately 3,000 species of larger fungi in the country the task is even more difficult. It would be folly therefore to indicate edibility for all the fungi described in a book such as this; the golden rule which should be adopted is not to eat any of the fungi one collects in the woods and fields. A fault of most popular treatments is that they are biased towards the human diet and selection of species is done on this basis; in the present work selection of examples within the 270 pages has been difficult and two factors have been particularly considered to ensure that (i) representatives of all the major groups of fungi and genera have been covered and (ii) a coverage has been attempted of all the common ecological niches.

I am fully aware that the taste of a fungus may be distinctive to that species or to a group of closely related species, but it is only a spot character and the tasting of one's finds is neither necessary nor advisable; indeed it is not used in this book. The odour, however, has been indicated whenever distinctive.

CONTENTS

Cover transparency supplied by John Markham, F.R.P.S., F.Z.S.

INTRODUCTION

The term larger fungus refers to any fungus whose study does not necessarily require more than a low-powered lens to see most of the important morphological features. Using such a term cuts across the existing scientific classification, for it includes the more obvious fungi bearing their spores on specialised reproductive cells called basidia, fig. 5, and a few of those whose spores are produced inside specialised reproductive cells called asci. The term is useful, however, even though it embraces a whole host of unrelated groups of fungi; it includes the polypores, fairy-clubs, hedgehog-fungi, puff-balls and elf-cups, as well as the more familiar mushrooms and toadstools—or puddockstools as they are often called in Scotland. Specimens of all these groups will find their way some time into the collecting baskets of the naturalist when he is out fungus-picking, along with probably a few jelly-fungi and less frequently one or two species of the rather more distantly related group, the morels. The biggest proportion of the finds, however, on any one collecting day in the autumn, when the larger fungi are in their greatest numbers, will be of the mushrooms and toadstools; these are, collectively, more correctly called the agarics.

The early botanists and pioneer mycologists of the nineteenth century recognised the fact that the fungi both large and small are ecologically connected to the herbaceous plants and trees among which they grow, but many mycologists since have tended to neglect these early observations. Although the importance of the fungi in the economy of the woodland, copse, field and marsh is well-known, mycologists and ecologists alike have been rather slow to appreciate that the fungi can be just as good indicators of soil conditions, if not better, than many other plants. Perhaps it is rash to attempt such a treatment as you find here because we know so little of the reasons why a particular fungus prefers one habitat to another. However, it is envisaged and hoped that, if a framework is provided, accurate field-notes can gradually be accumulated and many of the secrets yet to be uncovered explained.

Where to look
Fungi can be found in most situations which are damp at some time of the year. Searching for fungi can begin as soon as the spring days become warm, although even in the colder periods of winter several

finds can be made. In summer it gets very dry and this necessitates collecting in damper areas, such as marshes, alder-carrs, swamps and moorland bogs. After a heavy storm in summer, on the edges of paths and roadsides, woodland banks, in clearings in woods and in gardens, fungi can be collected within a few days of the rain, but collecting normally reaches a climax in August–September, the precise date depending on the locality and the individual character of the particular year.

All woodlands are worth visiting, particularly well-established woods with a mixture of trees. Pure pine-woods do not seem to be as good as pine-woods with scattered birch; plantations are often disappointing except after heavy rain or late in the season, even well into November in mild years. Pure birch and beech, the latter particularly when on chalky soils, are excellent areas to visit. Oak is possibly not as good but areas with willow and alder have many unique species. The edge of woods, sides of paths or clearings are usually more productive areas to search in than is the depth of the wood, and a small plot of trees can be much more rewarding than a large expanse of woodland. After some time one is able to judge the sort of place which will yield fungi. Rotten and burnt wood are very suitable substrates for they retain the moisture necessary for growth of fungi even in dry conditions, so allowing fructification to take place.

Grasslands including hill-pastures, established sand-dunes, etc., are often excellent, but of course they are much more dependent on the weather to produce favourable conditions for fungal development than woodland areas where the changes in the humidity and temperature are less extreme; prolonged mist or mild showery weather favour the fruiting of the grassland fungi. Dung in both woods and fields is an excellent although ephemeral substrate; many species of fungi characterise dung whilst others will grow in manured fields, on straw-heaps or where man has distributed the habitat.

Collecting

The collecting of larger fungi should not be considered a haphazard pursuit; careless collecting often results in many frustrating hours being spent on the identification of inadequate material, which is also not suitable after for preservation as reference material. A few good specimens are infinitely better than several poor ones; one is always tempted to collect too much and then collections are inevitably discarded. Always try to select specimens showing all the possible stages of development from the smallest buttons to the expanded caps. Sometimes such a range is not possible and one must be satisfied with either a couple or only one fruit-body.

Carefully dig up or cut from the substrate the entire fungus and handle it as little as possible. A strong pen-knife or fern-trowel is admirable for the job. The associated plants should be noted, especially trees, and if one is unable to identify the plants or woody debris retain a leaf or a piece of wood for later identification. One should note in a field-notebook any features which strike one as of interest, such as smell, colour, changes on bruising, presence of a hairy or viscid surface.

For transporting home the specimens should be placed in tubes, tins or waxed paper which are themselves kept in a basket. The smallest specimen can go in the first, the intermediate-sized forms in the tins or waxed paper and the larger ones laid in the basket or placed in large paper bags; plastic bags are not suitable except for very woody fungi. Thus an assortment of tins, tubes and various sizes of pieces of waxed paper are essential before setting out on a collecting trip. The specimens should be placed in the waxed paper such that they can be wrapped once or twice and the ends twisted as if wrapping a sweet.

Examination

Once home always aim at examining the specimens methodically.

The first necessity is to determine whether the fungus, which has been collected, has its spores borne inside a specialised reproductive cell (ascus) i.e. Ascomycete, or on a reproductive cell (basidium) i.e. Basidiomycete. By taking a small piece of the spore-bearing tissue, mounting in water, gently tapping it and examining under a low power of the microscope this can be easily ascertained. The tapping out is best done with the clean eraser of a rubber-topped pencil. There are several different shaped asci and basidia; the latter structures are more important in our study because the Ascomycetes are in the main composed of microscopic members.

The following procedure is necessary for the examination of your find:—

Select a mature cap of an agaric from each collection, cut off the stem and set the cap gills down on white paper, or if the specimen is small or is a woody or toothed fungus, or consists of a club or flattened irregular plate, place the spore-bearing surface (hymenium) face down on a microscope glass slide. The smaller specimens must be placed in tins with a drop of water on the cap to prevent drying out. Even with the larger specimens it is desirable to place a glass slide somewhere under the cap between the gills and the paper, and if possible to enclose the species carefully in waxed paper or in a tin. Whilst you are waiting for the spore-print to form, notes must be made on the more easily observable features; one is not required at this stage to examine the microscopic characters.

All the characters which may change on drying must be noted immediately, and these include colour, stickiness, shape, smell and texture. A sketch, preferably in colour, however rough, can give much more information than many score words.

Cut one fruit-body, longitudinally down with a razor or scalpel or a sharp knife if the fruit-body is woody, and sketch the cut surfaces, fig. 1A–B. These sketches and the rest of the collection notes should be made such that identification and future comparisons can be achieved. Thus always note the characters in the same order for each description. A table of the important characters is provided here, but this is meant as a guide not as a questionnaire. The attachment of the gills, pores or teeth to the fruit-bodies when once the fungus is in section should be always noted (see p. 20).

The spore-print when complete should be allowed to dry under normal conditions and then the spore-mass scraped together into a small pile. A microscope cover-slip should be placed on the top of the pile and lightly pressed down. The colour of the spore-print (or deposit) can then be compared with a standard colour chart and the spores making up the print examined in water under a microscope.

Microscopic examination

When one is more experienced with fungi it will be found necessary to carry out many microscopic observations, but when commencing the study it is necessary only to have an ordinary microscope; a calibrated eyepiece-micrometer is an advantage as is an oil-immersion lens. An examination of the spores is always necessary; the examination of features such as the sterile cells on the gill and stem, etc., varies with the fungus under observation. Spores should if at all possible be taken from a spore-print and mounted on a microscope slide, either in water or in a dilute aqueous solution of household ammonia. Although for mycologists it is often necessary to measure spores to within a $\frac{1}{2}$ micron (μm) this book has been so arranged that one only really has to distinguish between a spore which is small (up to 5 μm), medium (5–10 μm), long (10–15 μm), or large if globose and very long (if over 15 μm); this is

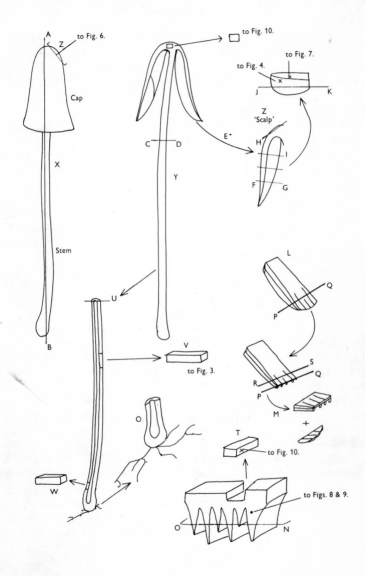

to Fig. 6.

to Fig. 10.

to Fig. 7.

to Fig. 4.

Cap

Z 'Scalp'

X

Y

Stem

to Fig. 3.

to Fig. 10.

to Figs. 8 & 9.

Fig. 1. Dissection of a toadstool as recommended by the author. For explanation see text.

not strictly accurate, but serves the purpose for an introductory text. It is important to describe the character of the spore, i.e. ornamentations, whether a hole (germ-pore) is present at one end and/or a beak (apiculus) at the other (fig. 5). With white or pale coloured spores it is useful to stain either the spore or the surrounding liquid with a dye—10% cotton blue solution is admirable, or a solution of 1·5 g iodine in 100 ml of an aqueous mixture containing 5 g of potassium iodine and 100 g of chloral hydrate. Both these dyes must be accurately made up if the study of the fungi is to be taken at all seriously; because some of the chemicals used above are not normally required by students, a chemist must make up the reagents for you. Often the spores turn entirely or partially blue-black or pale blue or purplish red in the iodine solution— a useful character.

Material in good condition is always required and one of the first things the student needs to do is train himself to collect sufficient material in good condition. The steps by which all the structures of the fungus used in the text can be observed are outlined below:—

Fig. 1 shows the cuts required to furnish suitable sections in order to observe the various structures and patterns of tissue which are important.

1. Carefully place the longitudinal section (AB) of the fruit-body which has been sketched gill-face down under a low power or dissecting microscope. Hairs or gluten on the cap, if present, will be made visible by focusing up and down (figs. 2 and 3A) and/or those on the stem (fig. 3B). When any part of the cut fruit-body is not being examined retain it in a chamber containing damp paper or moist moss; this will assist the cells to retain their turgidity, for they frequently collapse on drying and are difficult to observe except after performing often lengthy and special techniques.

If only one fruit-body is available, then cut along CD and mount in a tin box on a slide in order to obtain a spore-print (otherwise see paragraph 6).

2. Cut off a complete gill (E) and quickly mount on a dry slide. Under the low power of a microscope, the cystidia on the gill-margin will be visible (fig. 4); it will be seen whether the spores are arranged in a particular pattern (fig. 5) and whether the basidia are 2-spored or 4-spored. In white-spored toadstools it is difficult sometimes to determine whether the basidia are 2- or 4-spored so one must confirm the observations by other techniques.

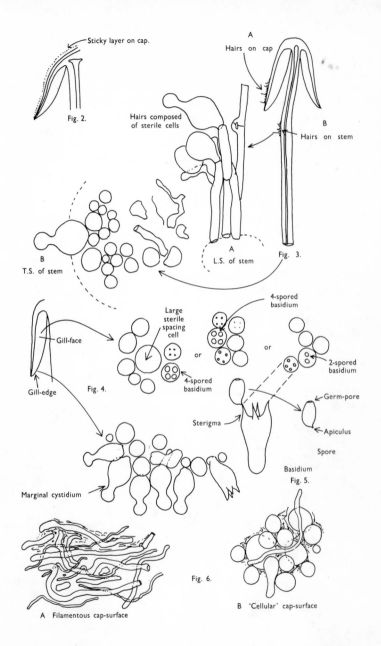

Sticky layer on cap.

Fig. 2.

A

Hairs on cap

B

Hairs composed of sterile cells

Hairs on stem

A
L.S. of stem

Fig. 3.

B
T.S. of stem

Gill-face

Large sterile spacing cell

4-spored basidium

or

4-spored basidium

or

2-spored basidium

Germ-pore

Gill-edge

Fig. 4.

4-spored basidium

Sterigma

Apiculus

Spore

Marginal cystidium

Basidium

Fig. 5.

A Filamentous cap-surface

Fig. 6.

B 'Cellular' cap-surface

A section of the gill accompanied by a small piece of cap-tissue, as in E, will confirm the presence or absence of noticeable cystidia (or hairs) on the cap. Now mount the section bounded by FG and HI in a drop of water containing either a drop of washing-up liquid and/or glycerine; the soapy liquid helps to expel any water which may tend to cling to the gill-margin amongst the cystidia and the glycerine stops the mount from drying out whilst further sections for comparison are cut and examined. It is at this time that the structure of the outermost layer of the cap can be examined, e.g. whether it is made up of a turf-like structure; the presence or absence of cystidia on the cap can be also confirmed (fig. 7A–C). It is frequently necessary to tap the mount in order to spread the tissue slightly and expose the elements; this can be done very efficiently by light pressure from the end of a pencil to which an eraser is attached. Cut off along line JK to eliminate marginal cystidia from confusing the picture and mount both pieces separately.
3. Cut out a wedge of tissue from the fruit-body (L) so as to have several gills attached to some cap-tissue; until one is familiar with the variability of facial and marginal cystidia, carefully cut along the line PQ (note: the cut is made one-third of the distance from the cap margin, thus eliminating the possibility of large numbers of marginal cystidia being examined in error for facial cystidia). Now make a second cut along the line of RS so that finally a small block of tissue remains (M).

Mount on a dry slide with the plane through PQ face down on the slide and observe under a low magnification, to assess whether cystidia on the gill-face are present or absent, and if present their general shape and whether numerous or infrequent (fig. 8).

Mount in water/washing-up mixture as outlined above and tap gently with the rubber attached to the end of a pencil; evenly distributed pressure should be given. If the gills appear to be too close then rotate the rubber a little whilst pressing in order to spread the tissue.
4. Using a low power of a microscope and looking down into the plane RS of the unmodified block M or a similar block, one obtains by this simple technique a very accurate idea as to the structure of the trama of the gill (fig. 9). The organisation of this tissue is very important in classification, some groups of toadstools having what has been described as regular trama (fig. 9C), others irregular (fig. 9D), inverse (fig. 9B) or divergent (fig. 9A). This same tissue may be thick or sparse to

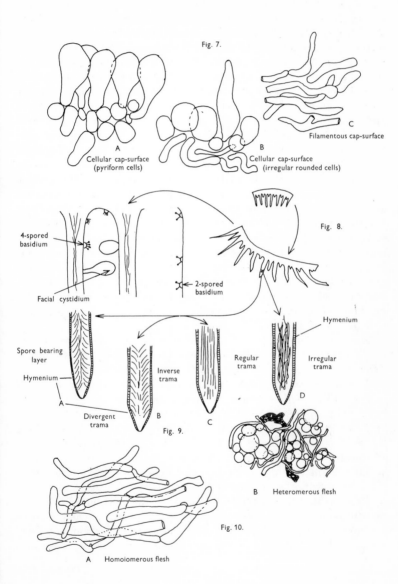

Fig. 7.

Cellular cap-surface
(pyriform cells)

A

Cellular cap-surface
(irregular rounded cells)

B

Filamentous cap-surface

C

Fig. 8.

4-spored basidium

Facial cystidium

2-spored basidium

Spore bearing layer

Hymenium

A

Divergent trama

Inverse trama

B

Regular trama

C

Hymenium

Irregular trama

D

Fig. 9.

B   Heteromerous flesh

A   Homoiomerous flesh

Fig. 10.

wanting, coloured or not. Such sections are often better than attempts at very thin sections unless very specialised techniques are used. There are few satisfactory thicknesses between the two extremes; the thick sections you can do and the very thin requiring expert techniques.

5. Take out a small block of tissue T as indicated in the figure (fig. 1). Mount immediately and repeat as in 3. This will allow the outer layer of the cap to be more clearly seen (fig. 7A–C) and also the structure of the flesh (fig. 10). The latter may be composed of a mixture of filaments and 'packets' or 'nests' of rounded cells (i.e. heteromerous), or of filaments, only some of which may be inflated (i.e. homoiomerous); but when individual cells are swollen they never form distinct groups. By very similar techniques it is possible to show that the more woody fungi can have flesh composed of one of four types of cells (Corner, 1932): these types depend on whether distinctly thickened cells (plate 47) are present with the actively growing hyphae or not (pp. 140–150), whether hyphae are present which bind groups of hyphae together, etc. (plate 46).

6. Remove stem along line CD and cut out small blocks of tissue as indicated (U, V and W). Mount immediately and examine as in paragraph 3, for cystidia, etc. (see fig. 3).

Whilst all these sections are being cut and processed a second fruit-body, if available, should be set to drop spores; this is done by cutting off the cap from the stem and placing it either entirely or in part, and with gill-edges down, on a slide in a tin.

7. Z is a 'scalp' of a cap; a thin sliver from the cap is placed on a slide in a drop of water (modified with washing-up liquid, etc. as above). After placing a cover-slip over the tissue it is tapped gently; this will show if the cap is composed of globose to elliptic elements or if it is composed of strictly filamentous units (figs. 6A & B). Care must be taken not to reverse the section when transferring it to the mountant, either by turning the scalpel or by allowing the surface tension of the liquid to pull the section upside down. The construction of any veil fragments will also be seen in this mount, and if a loose covering of veil is present this should be removed before observation so that it does not obscure the fundamental structures.

8. Examine the stipe of the fruit-body used above under a low power or with a dissecting microscope in order to ascertain whether there are any remains of veil and/or vegetative mycelium. If found, mount immediately in the solution containing iodine mentioned above and examine.

Of course it is difficult to carry out the above system the first time and be successful in seeing everything, indeed in being able to cut all the sections 1–8. Practice makes perfect, so why not practise with a $\frac{1}{4}$ lb of mushrooms from the grocer before the autumn season starts. In this way you will have overcome the difficulties without having to experiment with your collections.

CHARACTERISTICS FOR THE IDENTIFICATION OF HIGHER FUNGI WITH CAPS
Locality   G. Ref.   Date
Habitat notes   soil type   pH
                            vegetational community
                            solitary; in troops or rings
Draw or preferably paint exterior and vertical section of fruit-body
MACROSCOPIC CHARACTERS
CAP
    General characters:
        diameter   shape   consistency
        colour: when immature   when mature
                                when wet   when dry
    Surface
        dry, moist, greasy, viscid, glutinous, peeling easily or not,
        smooth, matt, polished, irregularly roughened, downy, velvety, scaly, shaggy
    Margin
        regular, wavy   incurved or not
        smooth, rough, furrowed   striate or not
    Veil, if present
        colour   abundance or scarcity
        distribution at margin, whether appendiculate or dentate
        consistency, whether filamentous, membranous
GILLS, or pores or teeth etc.
    remote, free, adnate, adnexed, emarginate, subdecurrent, decurrent
    crowded or distant   distinctly formed or not
    shape   interveined or not
    easily separable from the cap-tissue or not
    consistency (whether brittle, pliable, fleshy or waxy)
    thickness   width
    colour: when immature   at maturity
    number of different lengths or number of layers
    obvious features of gill-edge, tube-edge, e.g. colour, consistency
STEM
    central, eccentric or lacking   shape
    dimensions: length   thickness
    hollow or not
    colour: when immature   when mature
    consistency (whether fleshy, stringy, cartilaginous, leathery or woody)
    surface characters (whether fibrillose, dry, viscid, scaly or smooth)
    characters of stem-base
    Veil, if present   characters
    Volva, if present   characters
    Ring, if present
        whether single or double   whether membranous or filamentous
        whether persistent, fugacious or mobile whether thick or thin
        whether apical, median or basal
FLESH
    colour in cap: when wet   when dry
    colour in stem: when wet   when dry
    colour changes if any when exposed to air
    presence or absence of milk-like or coloured fluid
    (note: colour when exuded on fruit-body immediately and after some time
    and when dabbed on to a clean cloth or paper handkerchief and exposed to the air).
SMELL   before and after cutting   —relate to a common every day odour
MICROSCOPIC CHARACTERS
BASIDIOSPORES
    colour in mass   *   colour under microscope
    shape   size   type of ornamentation, if any
    size and shape of germ-pore, if present
    iodine reaction of spore-mass:—blue-black to dark violet (amyloid); red-purple
    (dextrinoid); yellow-brown or brown (non-amyloid)
BASIDIA   number of sterigmata
CAP-FLESH   type of constituent cells
GILL-TISSUE   type and arrangement of cells between adjacent hymenial faces
CAP-SURFACE   type of cells composing the outermost layer—whether filaments or
                     rounded cells
STERILE CELLS—CYSTIDIA
    presence or absence of sterile cells:—
    on gill-edge   on gill-margin
    on cap   on stem
    shape, estimation of size, thick or thin-walled, hyaline or not
    types of ornamentation, etc.

Key to the major classes of Larger Fungi

Spores borne externally on stalks on a clavate to cylindrical cell
　　　　　　　　　　　　　　　　　　　　　Basidiomycotina
Spores produced within a clavate, cylindrical or subglobose cell
　　　　　　　　　　　　　　　　　　　　　Ascomycotina

Key to major groups based on character of basidium and fruit-body shape

1. Basidia either produced in a hymenium or in a mass, and until
   maturity contained within a closed fruit-body .. 　Gasteromycetes
   Basidia produced in a layer of cells (hymenium) and exposed
   to the air before the maturity of the spores (Hymenomycetes)　2
2. Basidia simple, a single cell (fig. 5) (Homobasidiae) .. 　.. 　3
   Basidia usually septate, or if simple then fruit-body gelatinous
   and often collapsing to form a skin when dried (Heterobasidiae)　4
3. Fruit-body usually fleshy, soft and easily decaying (putrescent),
   hymenium spread over the surface of gills, ridges or within
   tubes 　.. 　.. 　.. 　.. 　.. 　Agaricales (p. 22)
   Fruit-body with hymenium smooth or spread-out on teeth,
   ridges or plates or if within tubes then fruit-body tough and
   leathery 　.. 　.. 　.. 　.. 　.. Aphyllophorales (p. 135)
4. Basidia divided 　.. 　.. 　.. 　.. 　.. 　.. 　.. 　5
   Basidia simple and apex drawn out into two long necks Plate
   61 (p. 185) 　.. 　.. 　.. 　Dacrymycetales (p. 180)
5. Basidia divided transversely by one to three horizontal septae
   Plate 60 (p. 183) .. 　.. 　.. 　Auriculariales (p. 182)
   Basidia divided into two or four cells by vertical septae Plate 61
   (p. 185) 　.. 　.. 　.. 　.. 　Tremellales (p. 184)

A. AGARICS AND THEIR RELATIVES

Key to major genera

1. Spores distinctly coloured in mass and coloured individually
   under the microscope 2
   Spores not, or faintly, coloured in mass and hyaline under the
   microscope 24
2. Spores blackish or some shade of brown 8
   Spores pinkish 3
3. Stem laterally attached to the cap or absent
         Claudopus (and some species of *Clitopilus*)
   Stem centrally attached to the cap 4
4. Stem with a cup-like structure enveloping the base   *Volvariella*
   Stem lacking any special structure at its base 5
5. Gills not attached to the stem (free), or with part attached to
   and descending down the stem (decurrent) 6
   Gills attached to the stem but not descending down the stem   7
6. Gills remote to free from the stem .. .. .. *Pluteus*
   Gills distinctly attached and descending down the stem
         Clitopilus (see also *Eccilia* p. 102)
7. Gills broadly attached to the stem (adnate) .. .. *Entoloma*
   Gills narrowly attached to the stem (adnexed) *Leptonia* & *Nolanea*
8. Stem laterally attached to the cap .. .. .. *Crepidotus*
   Stem centrally attached to the cap 9
9. Spore-print some shade of brown 10
   Spore-print blackish to purplish black 18
10. Spore-print bright rust-brown 11
    Spore-print dull clay-brown or ochraceous 16
11. Stem with the veil girdling the stem (ring), or cobweb-like
    (cortina) 12
    Stem without the veil girdling the stem or when present then
    easily lost 13

12. Stem with distinct ring or ring-zone   *Pholiota* & related genera
     Stem with cobweb-like veil or faint filamentous ring-zone
                                                 Cortinarius & *Gymnopilus*

13. Gills attached to the stem but not descending down the stem
     (adnexed to adnate)   14
     Gills free of the stem, or distinctly attached to and running
     down the stem (decurrent), and then often joined together
     at the apex of the stem or at their base   15

14. Cap-surface composed of rounded cells   ..   ..   *Conocybe*
     Cap-surface composed of filamentous cells   ..   ..   *Galerina*

15. Gills free of the stem and the whole fruit-body very fragile
                                                      Bolbitius

     Gills attached to and running down the stem (decurrent),
     easily separable from the cap-tissue and frequently veined at
     apex of stem   ..   ..   *Paxillus*

16. Cap scaly, fibrillose and roughened   ..   ..   ..   *Inocybe*
     Cap smooth, greasy or viscid   17

17. Cap-surface composed of rounded cells   ..   ..   *Agrocybe*
     Cap-surface composed of filamentous cells   *Naucoria* & *Hebeloma*

18. Gills or complete fruit-body becoming liquefied   ..   *Coprinus*
     Neither the gills nor fruit-body collapsing into a slurry of cells   19

19. Gills free to remote from the stem or attached and descending
     down the stem (decurrent)   20
     Gills attached in some way to the stem but not descending
     down the stem (adnate to adnexed)   21

20. Gills decurrent; stem possessing a cobweb-like veil
                               Gomphidius and *Chroogomphus*
     Gills remote or free; stem possessing a usually persistent ring
                                                 Agaricus

21. Gills distinctly spotted or distinctly mottled; stem stiff but
     breaking with a snap when bent; growing on dung or in
     richly manured areas   ..   ..   *Panaeolus*
     Gills not spotted or distinctly mottled; stem cartilaginous or
     not, and fruit-body growing on dung or not   22

22. Gills broadly attached to the stem (adnate) and with a veil
     girdling the stem   ..   ..   *Stropharia*
     Gills narrowly attached to the stem (adnexed) or with concave
     dentation near the stem (sinuate), or if adnate then lacking a
     ring   23

23. Gills with concave indentation near the stem (sinuate) and
cap and stem with a cobweb-like veil ..   ..   .. *Hypholoma*
Gills attached to the stem but lacking a distinct concave indentation near the stem   24

24. Stem stiff but breaking with a snap when bent; edge of cap
incurved at first and cap-surface composed of filamentous
cells   ..   .. *Psilocybe*
Stem fragile; edge of cap straight even when young and cap-
surface composed of rounded cells   ..   ..   .. *Psathyrella*

25. Fruit-body fleshy and readily decaying, often firm but never
tough   26
Fruit-body tough and not easily decaying   47

26. Parasitic on other agarics   ..   .. *Nyctalis*
Not parasitic on other agarics 27

27. Spore-bearing layer on fold-like often forked gills or simply
on irregularities 28
Spore-bearing layer (hymenium), on distinct well-formed gills 29

28. Spore-bearing layer on fold-like gills ..   .. *Cantharellus*
Spore-bearing layer on surface of irregularities   *Craterellus*

29. Cap easily separable from the stem 30
Cap not easily separable from the stem   31

30. Stem with girdling veil (ring) and/or with a persistent cup-
like structure at the base (volva); cap usually with warts or
scales distributed on its surface   ..   ..   .. *Amanita*
Stem with a ring but lacking a volva; cap surface powdery,
hairy or scaly   ..   ..   ..   *Lepiota* & related genera

31. Cap, stem and gills brittle; stem never stiff and either exuding
a milk-like juice or not; spores with spines or warts which
stain blue-black in solutions containing iodine   32
Cap, stem and gills soft or if stem stiff then snapping when
bent; gills never brittle 33

32. Fruit-body exuding a milk-like fluid ..   ..   .. *Lactarius*
Fruit-body not exuding milk-like fluid   ..   .. *Russula*

33. Gills thick, watery and lustrous (waxy) or with a bloom as if
powdered with talc; often brightly coloured 34
Gills not waxy and rarely over 1·5 mm thick   36

34. Gills rather watery and lustrous (waxy); spores smooth   .. 35
Gills rigid not watery, with powdery bloom; spores with
distinct spines   ..   .. *Laccaria*

35. Fruit-body with a distinct veil and growing in woods; cap
    often viscid or pale coloured   ..   ..   ..   *Hygrophorus*
    Fruit-body lacking a veil and usually growing in fields; cap
    usually brightly coloured and sometimes viscid   *Hygrocybe*
36. Stem with girdling veil (ring) and/or stem not attached to the
    centre of the cap (eccentric)   37
    Stem central and lacking a ring   38
37. Stem central and possessing a ring   ..   ..   .. *Armillaria*
    Stem not centrally attached to the cap

    members of the '*Pleurotaceae*' (p. 74)

38. Stem fibrous   39
    Stem stiff only in the outer layers   42
39. Gills with a concave indentation near the stem (sinuate)   ..   40
    Gills attached to and descending down the stem (decurrent)   41
40. Spores with warts which darken in solutions containing iodine

    Melanoleuca

    Spores not so colouring in solutions containing iodine

    Tricholoma & related genera

41. Spores with warts which darken in solutions containing iodine

    Leucopaxillus

    Spores not so colouring in solutions containing iodine

    Tricholoma & related genera

42. Gills thick and with rather blunt edges

    Cantharellula & *Hygrophoropsis*

    Gills thin and with distinct and sharp edges   43

43. Gills attached to and descending down the stem (decurrent);
    cap often depressed at the centre and sterile cells absent from
    the gills and the surface of the cap   ..   *Clitocybe* & *Omphalina*
    Gills attached to the stem but not descending down the stem
    (adnate to adnexed) or if descending then distinct sterile cells
    on the gills, cap and stem   44
44. Cap-edge straight and usually striate when young; cap thin and
    somewhat conical and gills descending down the stem or not

    Mycena & related genera

    Cap-edge incurved, non-striate and cap rather fleshy; gills
    not descending down the stem   45

45. Stem dark and woolly at least in the lower half and the cap
    viscid; fruit-bodies growing in clusters on tree-trunks

    Flammulina

Stem not dark and woolly 46

46. Cap viscid and stem usually rooting; fruit-body growing directly on wood or attached to wood by long strands or cords of mycelium (rhizomorphs) .. .. .. *Oudemansiella*

If cap viscid then fruit-body neither attached to wood by cords of mycelium nor stem with a rooting base

Collybia & related genera

47. Stem central and gills often interconnected by veins; cap can be dried and later revived, purely by moistening

Marasmius & related genera

Stem not attached to the centre of the cap and fruit-body although persistent not easily revived to natural shape after once being dried 48

48. Spore-print blue-black with solutions containing iodine .. 49

Spore-print yellowish in solutions containing iodine 50

49. Gills toothed or notched along the edges .. .. *Lentinellus*

Gills even along their edges and not toothed .. *Panellus*

50. Gills appearing as if split down their middles *Schizophyllum*

Gills not splitting 51

51. Gills notched or toothed along their edges .. .. *Lentinus*

Gills even along their edges and not toothed .. *Panus*

52. Spore-print yellowish, purplish, black or pink 53

Spore-print some shade of brown, but without purplish flush 56

53. Spore-print yellowish or pinkish 54

Spore-print purplish brown or blackish 55

54. Spore-print yellowish .. .. *Gyroporus*

Spore-print pinkish .. .. *Tylopilus*

55. Spore-print purplish brown .. .. .. *Porphyrellus*

Spore-print blackish and spores ornamented .. *Strobilomyces*

56. Cap glutinous and stem with or without girdling veil (ring); within the tubes the sterile cells (cystidia) cluster together *Suillus*

Cap at most viscid and then only in wet weather and sterile cells within the tubes individually placed 57

57. Stem-surface covered with distinct black or dark brown or white then darkening scales; spore-print clay-brown with or without a flush of cinnamon-pinkish brown .. .. *Leccinum*

Stem-surface covered completely or in part with a network or pattern of faint lines or pale yellow or red-rust but never black dots; spore-print olivaceous buff *Boletus* & related genera

(i) Agarics of woodlands and copses

(a) Mycorrhizal formers

Leccinum scabrum (Fries) S. F. Gray

Birch rough stalks or Brown birch-bolete.

Cap: width 45–150 mm. *Stem:* width 70–200 mm; length 20–30 mm.

Description: Plate 1.

Cap: convex and becoming only slightly expanded at maturity, pale brown, tan or buff, soft, surface dry, but in wet weather becoming quite tacky, smooth or streaky-wrinkled and cap-margin not over-hanging the tubes.

Stem: white, buff or greyish, roughened by scurfy scales which are minute, pale and arranged in irregular lines at the stem-apex, and enlarged and dark brown to blackish towards the base.

Tubes: depressed about the stem, white becoming yellowish brown at maturity, with small, white pores which become buff at maturity and bruise distinctly yellow-brown or pale pinkish brown when touched.

Flesh: watery, very soft in the cap lacking distinctive smell and either not changing on exposure to the air or only faintly becoming pinkish or pale peach-colour.

Spore-print: brown with flush of pinkish brown when freshly prepared.

Spores: very long, spindle-shaped, smooth, pale honey-coloured under the microscope and more than 14 μm in length (14–20 μm long × 5–6 μm broad).

Marginal cystidia: numerous and flask-shaped. Facial cystidia: sparse, similar to marginal cystidia.

Habitat & *Distribution:* Found in copses and woods containing birch trees, or even accompanying solitary birches.

General Information: This fungus is recognised by the pale brown cap, the white, unchanging or hardly changing flesh and the cap-margin not overhanging the tubes. There are several closely related fungi which also grow with birch trees but they need some experience in order to distinguish them. This fungus was formerly placed in the genus *Boletus*, indeed it will be found in many books under this name. Species of *Leccinum* are edible and considered delicacies in continental Europe. The majority can be separated from the other fleshy fungi with pores beneath the cap, i.e. boletes, by the black to brown scaly stem and rather long, elongate spores. The scales on the stem give rise to the common name 'Rough stalks' which is applied to this whole group of fungi.

Illustrations: F 39C; Hvass 253; LH 122; NB 155[6]; WD 89[1].

Suillus grevillei (Klotzsch) Singer   Larch-bolete

Cap: width 30–100 mm. *Stem:* width 15–20 mm; length 50–70 mm.

Description: Plate 2.

Cap: convex or umbonate at first, later expanding and then becoming plano-convex, golden-yellow or rich orange-brown, very slimy because of the presence of a pale yellow sticky fluid.

Stem: apex reddish and dotted or ornamented with a fine network, cream-coloured about the centre because of the presence of a ring which soon collapses, ultimately appearing only as a pale yellow zone; below the ring the stem is yellowish or rusty brown, particularly when roughly handled.

Tubes: adnate to decurrent, deep yellow but becoming flushed wine-coloured on exposure to the air, with angular and small sulphur-yellow pores which become pale pinkish brown to lilaceous or pale wine-coloured when handled.

Flesh: with no distinctive smell, pale yellow immediately flushing lilaceous when exposed to the air, but finally becoming dingy red-brown, sometimes blue or green in the stem-base.

Spore-print: brown with distinct yellowish tint when freshly prepared.

Spores: long, ellipsoid, smooth and pale honey when under the microscope, less than 12 μm in length (8–11 μm long × 3–4 μm broad).

Marginal cystidia: in bundles and encrusted with amorphous brown, oily material. Facial cystidia: similar in shape and morphology to marginal cystidia.

Habitat & *Distribution:* Found on the ground accompanying larch trees either singly or more often in rings or troops.

General Information: This fungus is easily recognised by the poorly developed ring, overall golden-yellow colour and pale yellow viscidness on the cap which comes off on to the fingers when the fruit-

Plate 1. Fleshy fungi: Spores borne within tubes

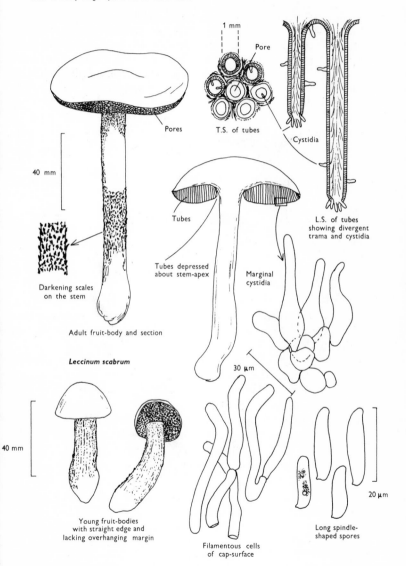

1 mm

Pore

Pores

T.S. of tubes

Cystidia

40 mm

Darkening scales
on the stem

Adult fruit-body and section

Leccinum scabrum

Tubes

Tubes depressed
about stem-apex

Marginal
cystidia

L.S. of tubes
showing divergent
trama and cystidia

40 mm

Young fruit-bodies
with straight edge and
lacking overhanging margin

30 μm

Filamentous cells
of cap-surface

Long spindle-
shaped spores

20 μm

Plate 2. Fleshy fungi: Spores borne within tubes

Suillus grevillei

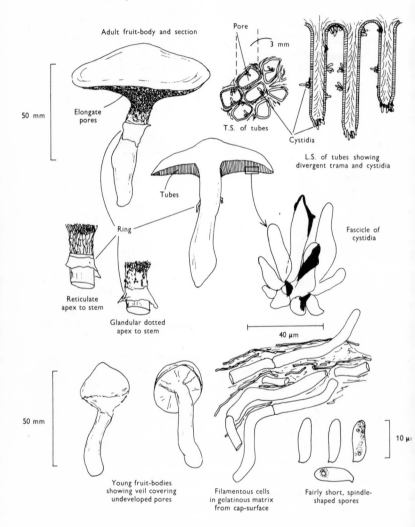

Adult fruit-body and section

50 mm

Elongate pores

Pore

3 mm

T.S. of tubes

Cystidia

L.S. of tubes showing divergent trama and cystidia

Tubes

Ring

Fascicle of cystidia

Reticulate apex to stem

Glandular dotted apex to stem

40 µm

50 mm

Young fruit-bodies showing veil covering undeveloped pores

Filamentous cells in gelatinous matrix from cap-surface

Fairly short, spindle-shaped spores

10 µ

body is handled. There are several closely related fungi which also grow with coniferous trees, e.g. *Suillus luteus* Fries, 'Slippery jack', but many need experience in order to identify them. All these fungi were formerly placed in the genus *Boletus*, because of the fleshy fruit-body with pores beneath the cap. The larch-bolete receives its common name from the close relationship of the fungus with the larch. On drying *S. luteus* and *S. grevillei* may strongly resemble one another but the former can be distinguished when fresh by the chocolate brown, sepia, or purplish brown cap and the large whitish, lilac-tinted ring.

Species of *Suillus* are edible and rank highly in continental cook-books, although they have disagreeably gelatinous-slimy caps, a character, in fact, which helps to separate them from other fleshy pore-fungi.

Illustrations: F 41a; Hvass 257; ML 187; NB 104[4]; WD 84[2].

Boletus badius Fries   Bay-coloured bolete

Cap: width 70–130 mm. *Stem:* width 34–37 mm; length 110–125 mm. (36–40 mm at base).

Description: Plate 3.

Cap: hemispherical, minutely velvety, but soon becoming smooth and distinctly viscid in wet weather, red-brown flushed with date-brown and darkening even more with age and in moist weather to become bay-brown.

Stem: similarly coloured to the cap but paler particularly at the apex, smooth or with faint, longitudinal furrows which are often powdered with minute, dark brown dots.

Tubes: adnate or depressed about the stem, lemon-yellow but immediately blue-green when exposed to the air and with angular, rather large similarly coloured, pores which equally rapidly turn blue-green when touched.

Flesh: strongly smelling earthy, pale yellow but becoming pinkish in centre of the cap, and blue in the stem and above the tubes when exposed to the air, but finally becoming dirty yellow throughout.

Spore-print: brown with a distinct olivaceous flush.

Spores: long, spindle-shaped, smooth, honey-coloured under the microscope and greater than 12 μm in length (13–15 μm long × 5 μm broad).

Marginal cystidia: numerous, flask-shaped and slightly yellowish.

Facial cystidia: scattered and infrequent and similar to marginal cystidia in shape.

Habitat & Distribution: Found in woods, especially accompanying pine trees, but often found fruiting on the site of former coniferous trees, even years after the trunks or the stumps have been removed.

General Information: This fungus is recognised by the rounded, red-brown cap, coupled with the pale yellow flesh and greenish yellow tubes, both of which become greenish blue when exposed to the air. There are several species in the genus *Boletus* which stain blue at the slightest touch or when the flesh is exposed to the air, e.g. *B. erythropus* (Fries) Secretan, a common bolete with a dark olivaceous cap, orange pores and red-dotted stem.

The flesh of some species of *Boletus*, e.g. *B. edulis* Fries, however, remains unchanged or at most becomes flushed slightly pinkish. Although many people say they recognise *B. edulis*, the 'Penny-bun' bolete—a name derived from the colour of the cap, there is some doubt as to whether the true *B. edulis* is common in Britain as we are led to believe. *B. edulis* and its relatives are highly recommended as edible (see p. 35). *B. badius* is also edible, but it is ill-advised to eat any bolete which turns blue when cut open.

Illustrations: B. badius—F 38c; Hvass 248 (not very good); LH 191; NB 109[5]; WD 85[1]. *B. edulis:* F 42a; Hvass 246; LH 191; NB 143[3].

General notes on Boletes

There are nearly seventy boletes recorded for the British Isles and evidence of others which have as yet not been fully documented. As a group they are characterised by being fleshy, possessing a central stem and producing their spores within the tubes, and not on gills as in the common mushroom. It is the first character by which the boletes differ so markedly from the other pored fungi, such as the 'Scaly Polypore' (see p. 140).

The boletes have long been classified in the genus *Boletus*, but instead of referring all the pored, fleshy fungi to a single large genus several genera are now recognised. The separation of these genera is based on differences in colour of the spore-print and differences in the anatomy of the tubes, cap and stem, etc., e.g. members of the genus

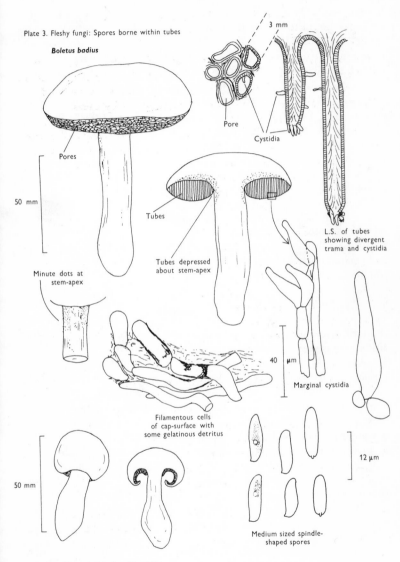

Plate 3. Fleshy fungi: Spores borne within tubes

Boletus badius

3 mm

Pore

Cystidia

Pores

50 mm

Tubes

Tubes depressed
about stem-apex

L.S. of tubes
showing divergent
trama and cystidia

Minute dots at
stem-apex

40 μm

Marginal cystidia

Filamentous cells
of cap-surface with
some gelatinous detritus

12 μm

50 mm

Medium sized spindle-
shaped spores

Young fruit-body with section
showing incurved margin

Suillus have colourless or pale coloured dots on the stem exuding a resin-like liquid in wet weather, which is clear and glistening in some species but turbid and whitish in others, gradually darkening and hardening so that the stem is ultimately covered in dark brown or reddish smears or spots; members of the genus *Leccinum* on the other hand never exude liquid and have coarse or fine roughenings on the stem which are usually dark, but may commence white and ultimately darken depending on the species; many species of *Boletus* possess a very distinct raised network all over the stem, whilst others have it present only in part, or have minute, often brightly coloured, dots replacing it.

Within this single, yet not particularly large, group of fungi, several biological phenomena are demonstrable. There is good evidence that the majority of British boletes are mycorrhizal; several species are known to be associated only with one species of tree or group of closely related tree-species. Thus *Suillus grevillei* and *S. aeruginascens* (Secretan) Singer grow in association with larch trees: *S. luteus* and *Boletus badius* in contrast grow in association with pine trees; *Leccinum scabrum* with birch trees; *L. aurantiacum* (Fries) S. F. Gray, with poplar trees and *L. quercinum* (Pilát) Green & Watling, with oak trees.

Boletus edulis can be separated into several distinct subspecies which are associated with different trees; the two commonest subspecies are those associated with birch and with beech trees. It is well known that although present in this country during the warmer periods of the Ice-Age, larch neither survived the intense cold of the last advance of the ice nor migrated back into Britain after the ice had melted. Thus all larches which we see in Britain have been planted by man. There is little doubt that mycelia of many fungi were introduced along with these plants very probably including the mycelium of the larch-bolete. A similar pattern can be seen with other introduced trees, although not to such a marked degree, e.g. spruce trees. The beech tree, however, is native to the south of England, unlike the larch returning to this country after the ice had melted; it has been planted extensively outside its former range in northern areas of the British Isles taking with it its associated fungi. There is some evidence that some stocks of beech and fungi have been introduced from continental Europe in comparatively recent times.

A parallel, yet inexplicable association is found between the bolete *Suillus bovinus* (Fries) O. Kuntze and its close relative *Gomphidius roseus* (Fries) Karsten where the mycelium of two fungi are found inter-twined forming a close association! Parasitism although rare is also found amongst the boletes, and an uncommon parasitism at that—a fungus on a fungus; for example in Britain although infrequent *Boletus parasiticus* Fries grows attached and ultimately replaces the spore-tissue of the common earth-ball (*Scleroderma*, see p. 192).

Those fungi which grow on dead and decaying substrates are called saprophytes and although the greater number of higher fungi would be included in this class of organisms the character is infrequent amongst the boletes. One British example of this type of fungus is the rare *Boletus sphaerocephalus* Barla which grows on woody debris.

Chemists have long been interested in boletes, for as noted above the flesh of some species when exposed to the atmosphere turns vivid colours, a feature often incorporated into the Latin name, e.g. *Boletus purpureus* Persoon, from the purple colours produced whenever the fruit-body is handled. The reaction appears to be an oxidation where in the presence of an enzyme and oxygen a pigmented substance or substances are produced. What the significance of these colour-changes is in nature is as yet unknown; however, what is interesting is that many of the chemicals involved are unique and have only recently been analysed completely; they are related to the quinones.

There is little doubt that it is this rapid and intense blueing of the flesh of many boletes that has lead to a belief that they are poisonous. It is uncertain whether there are any truly toxic species of *Boletus* but several have unpleasant smells and tastes which make them very unattractive. *Boletus edulis* is the important ingredient, however, which gives the distinctive taste to 'so-called' dried mushroom soup. Thou-sands of fruit-bodies are collected annually in the forests of Europe to be later dried and processed for incorporation into soup. Boletes appear to form an important part of the diet of several rodents and deer and in Scandinavia in the diet of reindeer.

Probably one of the most obscure of our British boletes is *Strobilo-myces floccopus* (Fries) Karsten, the 'Old Man of the Woods'. It has a black, white and grey woolly, scaly cap and stem, and the flesh dis-tinctly reddens when exposed to the air. The spores are almost spherical, purple-black in colour and covered in a coarse network when seen under the microscope. All these characters readily separate *Strobilo-myces* from all other European boletes; however, in Australasia, members of this and related genera form a very important part of the flora.

Chroogomphus rutilus (Fries) O. K. Miller   Pine spike-cap
Cap: width 30–150 mm. *Stem:* width 10–18 mm; length 60–120 mm.
Description:
Cap: convex with a pronounced often sharp umbo, wine-coloured,
    flushed with bronze-colour at centre and yellow or ochre at
    margin, viscid but soon drying and then becoming paler and quite
    shiny.
Stem: yellowish orange, apricot-coloured or peach-coloured, streaked
    with dull wine-colour, spindle-shaped or narrowed gradually to the
    apex from a more or less pointed base.
Gills: arcuate-decurrent, distant, at first greyish sepia then dingy
    purplish with paler margin, but finally entirely dark purplish brown.
Flesh: lacking distinctive smell and reddish yellow or pale tan in the
    cap, rich apricot- or peach-colour towards the stem-base.
Spore-print: purplish black.
Spores: very long, spindle-shaped, smooth, olivaceous purple and
    greater than 20 μm in length (20–23 × 6–7 μm).
Marginal cystidia: cylindrical to lance-shaped and up to 100 × 15 μm.
Facial cystidia: similar to marginal cystidia.
Habitat & Distribution: Found in pine woods, usually solitary or in
    small groups. Fairly common throughout the British Isles and
    characteristic of Scots Pine woods.
General Information: This fungus can be distinguished by the purplish
    or wine-coloured cap and the gills being pigmented from youth.
    There is only one other British species of this genus, i.e. *C. corallinus*
    Miller & Watling.

    Chroogomphus is separated from *Gomphidius* by the flesh having
an intense blue-black reaction when placed in solutions containing
iodine, and the gills being coloured from their youth. In many
books *Chroogomphus* is placed in synonymy with the genus *Gomphidius*. However, *Gomphidius glutinosus* (Fries) Fries, *G. roseus* (Fries)
Karsten and *G. maculatus* Fries all have whitish gills when immature
which gradually darken, and their flesh simply turns orange-brown
in solutions of iodine. *G. glutinosus* is uniformly grey in colour and
is most frequently found under spruce and other introduced conifers:
G. roseus has a pale-pinkish coloured cap and white stem, and grows
with pine; *G. maculatus* grows under larch and is flushed lilaceous
at first but becomes strongly spotted with brown when handled.
Illustrations: Hvass 192; LH 213; WD 83[3].

Plate 4. Fleshy fungi: Spores blackish and borne on gills

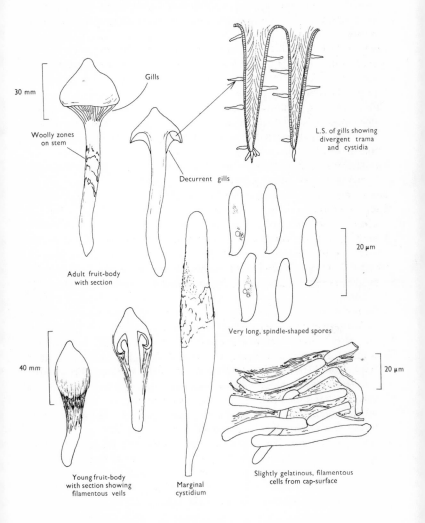

Gills

30 mm

Woolly zones
on stem

Decurrent gills

L.S. of gills showing
divergent trama
and cystidia

Adult fruit-body
with section

20 μm

Very long, spindle-shaped spores

40 mm

20 μm

Young fruit-body
with section showing
filamentous veils

Marginal
cystidium

Slightly gelatinous, filamentous
cells from cap-surface

Paxillus involutus (Fries) Karsten   Brown roll-rim
Cap: width 50–120 mm. *Stem:* width 8–15 mm; height 30–75 mm.
Description:
Cap: at first covex with a strongly inrolled, downy margin, but then
    expanded and later frequently depressed towards the centre, clay-
    coloured, ochre or yellow-rust, slightly velvety but becoming smooth
    or sticky particularly in wet weather and readily bruising red-brown
    when fresh.
Stem: central or slightly eccentric, thickened upwards, fibrillose-silky,
    similarly coloured to the cap but typically streaked with red-brown
    particularly with age.
Gills: ochre or yellow-brown then rust and finally darker brown,
    decurrent, crowded, often branched and united about the apex of
    the stem; easily peeled from the flesh with the fingers and rapidly
    becoming red-brown on handling.
Flesh: thick, soft and with slightly astringent smell and yellowish to
    brownish but becoming red-brown after exposure to the air.
Spore-print: rust-brown.
Spores: medium-sized, ellipsoid, smooth, deep yellow-brown and
    rarely greater than 10 μm in length (8–10 × 5–6 μm).
Marginal cystidia: numerous lance-shaped or spindle-shaped.
Facial cystidia: scattered and similar in shape to marginal cystidia.
Habitat & *Distribution:* Found on heaths and in mixed woods, parti-
    cularly where birch has or is now growing, or even accompanying
    solitary birch trees.
General Information: This fungus is easily recognisable by the strongly
    inrolled, woolly margin of the cap and yellow-brown gills which
    are easily separable from the cap-flesh. *P. rubicundulus* P. D. Orton
    is similar but grows under alder and has yellow gills unchanging when
    handled and dark scales on the cap. *P. atrotomentosus* (Fries) Fries
    and *P. panuoides* (Fries) Fries both grow on coniferous wood and
    have smaller spores; the former is recognised by the dark brown to
    almost black shaggy stem and the latter by the shell-shaped cap
    devoid almost completely of a stem.
Illustrations: F 41c; Hvass 189; LH 185; NB 115[8]; WD 70[2].

Plate 5. Fleshy fungi: Spores brown and borne on gills

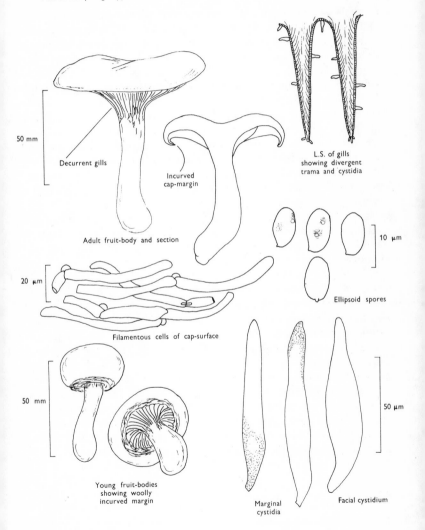

50 mm

Decurrent gills

Incurved
cap-margin

L.S. of gills
showing divergent
trama and cystidia

Adult fruit-body and section

10 μm

Ellipsoid spores

20 μm

Filamentous cells of cap-surface

50 mm

50 μm

Young fruit-bodies
showing woolly
incurved margin

Marginal
cystidia

Facial cystidium

Cortinarius pseudosalor J. Lange

Cap: width 60–125 mm. *Stem:* width 15–25 mm; length up to 180 mm.
Description:

Cap: bell-shaped or bluntly conical only slightly expanding with maturity, smooth or wrinkled at centre but often furrowed at the margin, slimy, brown with a distinct olive flush when in fresh condition and becoming ochraceous brown and shiny when dry.

Stem: usually swollen to some degree about the middle, slimy particularly towards the base, whitish throughout when young except for a faint amethyst or violaceous flush in the lower part; as the slime dries the stem becomes shiny and the outer surface breaks up into fibrillose scales or scaly, irregular ring-zones.

Flesh: lacking distinct smell, white with ochraceous flush in the cap, white in the stem, thick and soft in the cap but fibrous in the stem.

Gills: adnate, broad, rather thick, frequently veined and distant, ochraceous brown and finally deep rust-brown.

Spore-print: rust-colour.

Spores: long, slightly almond-shaped in side view, finely warted throughout and not less than 12 μm in length (13–14 × 7–8 μm).

Marginal cystidia: ellipsoid or club-shaped, hardly different from the surrounding undeveloped basidia.

Facial cystidia: absent.

Habitat & *Distribution:* Found on the ground in copses and woods especially those containing beech.

General Information: Recognised by the conical, grooved cap and the slimy spindle-shaped stem with a distinct violaceous flush; this fungus is often misnamed *C. elatior* Fries but this is a much less common fungus. There are several closely related fungi, but these grow with other tree-species and need much more experience to distinguish one from the other. *C. pinicola* P. D. Orton is one such species growing in the litter under *Pinus sylvestris*, Scots Pine; this species is fairly common in the remnant pine woods of Northern Scotland. The large size, sticky or glutinous cap and stem indicate that this fungus belongs to *Cortinarius*, subgenus *Myxacium*.

Illustrations: Hvass 145; LH 162; NB 119; WD 60[1].

Plate 6. Fleshy fungi: Spores brown and borne on gills

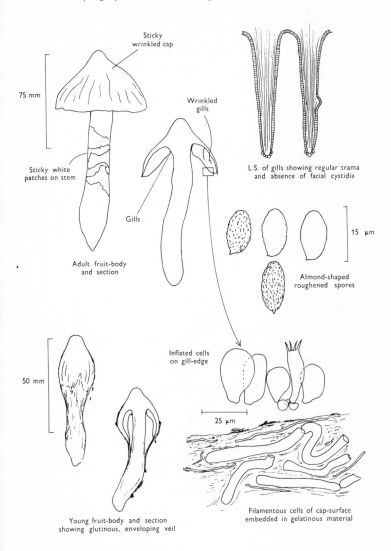

Sticky wrinkled cap

75 mm

Sticky white patches on stem

Adult fruit-body and section

Wrinkled gills

Gills

L.S. of gills showing regular trama and absence of facial cystidia

15 μm

Almond-shaped roughened spores

Inflated cells on gill-edge

25 μm

50 mm

Young fruit-body and section showing glutinous, enveloping veil

Filamentous cells of cap-surface embedded in gelatinous material

General notes on Cortinarii

The genus *Cortinarius* is the largest genus of agarics in the British Isles, indeed in Europe and North America—perhaps in the world. It includes some of our most beautiful agarics, yet it is one of the least satisfying to the mycologist because of the difficulties experienced in identifying collections—partly because many species are so seldom seen.

Cortinarius contains just under two hundred and fifty recognisable British species, although recent research has shown that many more are yet to be described from this country as new to science. Except for some very characteristic species the individual members within the genus *Cortinarius* are often very difficult to separate one from the other; however, *Cortinarius* is one of our least difficult genera to recognise in the field owing to the presence when mature of rust-coloured gills and a cobwebby veil which extends from the margin of the cap to the stem. This structure is termed a cortina (Fig. 14) and in young specimens covers the gills with delicate filaments. As the cap expands the cottony or cobwebby filaments are stretched and either disappear entirely or may collapse to form a ring-like zone of filaments on the stem. In some species a second completely enveloping veil is also found, and this veil is viscid in one distinct group of which *C. pseudosalor* already described is a member. The gills in the genus are variable in colour when young although constant for a single species; they may be lilaceous purple, orange, brown, red, yellow-ochraceous or tan, but ultimately in all members at maturity they become rust-colour. The spores under the microscope are richly coloured, yellow to red-brown and are frequently strongly warted; in mass they are rust-brown and this character coupled with the presence of the cobweb-like veil characterises the genus.

Within the genus *Cortinarius* there is a wide range of characters varying from species with distinctly sticky caps and stems, some with sticky caps and dry stems to those with both dry caps and stems. A few species are very large and fleshy whilst others are quite slender and many of the latter rapidly change colour on drying out and are then said to be hygrophanous. However, although there is such a large spectrum of characters in a single genus the species all have in common the cortina and rust-coloured gills, the latter often appearing as if powdered with rusty dust.

Utilising the characters mentioned above this very large genus can be split into the following six sections, called by the mycologist sub-genera

a. Large to medium sized fleshy agarics with viscid caps and stems— *Myxacium*

b. Large, fleshy agarics with viscid or tacky caps when fresh but dry stems—*Phlegmacium*

c. Large to medium sized agarics with dry, scaly or humid caps and dry stems which if orange tawny are robust—*Cortinarius*

d. Medium, rarely large, agarics with dry, silky to innately fibrillose caps, slender stems and frequently with at least part of the fruit-body yellow, orange or reddish—*Dermocybe*

e. Medium to small agarics with silky fibrillose, non-hygrophanous caps which may become tacky in wet weather and then usually with robust, clavate-bulbous stems—*Sericeocybe*

f. Small, less frequently medium or large agarics, all with distinctly hygrophanous caps—*Hydrocybe*.

In several continental books some or all of these divisions are recognised as distinct genera in their own right. The subgenus *Telamonia* which occurs in many texts was formerly thought to differ from *Hydrocybe* in the presence of a universal veil; the universal veil is a second veil which completely envelopes the fruit-body when it is young and is in addition to the cortina. However, the modern treatment would seem to suggest that the presence of the universal veil is not of the utmost importance and so the two subgenera are incorporated into one. The name *Hydrocybe* reflects the character of changing colour as it dries out because of the loss of water. Within each subgenus the species are distinguished by the colour of the young gills and of the cap, the veil colour and texture, and microscopic characters of the spores, particularly their size.

The majority of species of *Cortinarius* are mycorrhizal and like the boletes possess very specific relationships with tree species. Thus some are typical of coniferous woodland and others typical of deciduous woodland in general, whilst others typify woods of a particular tree, e.g. beech, oak, birch, pine, larch. Some species are characteristic of woods on limestone or chalky soils (calcareous) whilst others are characteristic of woods on sandy, heathy acidic soils. For example, *Cortinarius armillatus* (Fries) Fries which is found in damp woods and possesses one or more cinnabar-red or scarlet zones on the stem and red fibrils at the stem-base appears to be connected with birch. Several species are associated with native trees whilst others have undoubtedly been introduced from abroad. They are very important in the economy of the woodland ecosystem.

One of the most beautiful and easily distinguished of our British species is *Cortinarius violaceus* (Fries) Fries which has uniformly deep violet-coloured stem and cap and coloured cystidia on the gill-margin, a character unusual in *Cortinarius*.

No species are known to be truly poisonous and many species are known to be edible, but many are too small to be of any value. Some of the larger species are regarded as good to eat, but frequently are too scarce. Thus the necessity for experience to recognise the different species, coupled with their often unpleasant tastes make them an unimportant group of agarics for eating.

Russula ochroleuca (Secretan) Fries   Common yellow russula
Cap: width 50–100 mm. Stem: width 20–35 mm; length 50–100 mm.
Description: Plate 7.

Cap: yellow-ochre or dull yellow becoming paler with age, or flushed
    faintly greyish green, convex but soon expanding and becoming
    flat or depressed in the centre, smooth, or granular when young and
    slightly tacky in wet weather, faintly striate at the margin.

Stem: white at first then flushed slightly greyish, smooth or wrinkled,
    firm at first but quickly becoming soft and fragile.

Flesh: brittle, firm at first then soft, white, yellow under cap-centre.

Gills: white at first then flushed pale cream-colour, brittle, adnexed
    to free, rather distant.

Spore-print: faintly cream when freshly prepared.

Spores: medium-sized, hyaline, broadly ellipsoid or subglobose to
    almost globose, coarsely ornamented with prominent warts which
    stain blue-black when mounted in solutions containing iodine and
    which are faintly interconnected by low ridges, about 8×7 μm in
    size ($9–10 \times 7–8$ μm).

Marginal cystidia: prominent, lance- to spindle-shaped and often filled
    with oily material.

Facial cystidia: similar in shape to marginal cystidia and projecting
    some distance from the gill-face.

Habitat & *Distribution:* Commonly found in mixed woods from summer
    until late autumn.

General Information: Easily recognised by the ochre-yellow cap, very
    pale cream-coloured spore-print and greying stem. Two other
    yellow-capped species of *Russula* are commonly found. *R. claroflava*
    Grove with yellow spore-print and blackening fruit-body which grows
    with birches in boggy places, and *R. lutea* (Fries) S. F. Gray which is
    much smaller, having a cap up to 50 mm and very deep egg-yellow
    gills and spore-print; it grows in deciduous woods.

Illustrations: F 22a; Hvass 226; LH 119; NB 137[1]; WD 49[1].

General notes on the genus *Russula*

A large genus with nearly one hundred distinct species in the British
Isles and several others yet unrecognised or undocumented. This genus
is composed generally of large toadstools often beautifully coloured,
indeed the majority have brightly coloured caps in reds, purples,

yellows or greens depending on the species although a few are pre-dominantly white bruising reddish brown or grey to some degree.

Such large and distinctive fungi one would think would be the easiest members of our flora to identify, unfortunately they are not. They form a group quite isolated in their relations, the only close relatives being members of the genus *Lactarius*, to be dealt with later (see p. 50). The flesh of members of both *Lactarius* and *Russula* contains groups of rounded cells, a feature unique amongst agarics and explains why in *Russula* the fruit-bodies, cap and gills and some-times the stem are brittle and easily break if crushed between the fingers. The fruit-body does not exude a milky liquid when the flesh is broken.

The spore-print varies, depending on the species involved, from white to deep ochre and individual spores are covered in a coarse ornamenta-tion which is composed of isolated warts or warts interconnected by raised lines, or mixtures of both. The ornamentation stains deep blue-black when the spores are mounted in solutions containing iodine and the pattern which is produced appears in many cases to be of a specific character.

The majority of the species, if not all north-temperate species are mycorrhizal and the familiar host-tree fungus relationship can be recognised:—

R. claroflava Grove, with birch in boggy places, *R. emetica* (Fries) S. F. Gray with pine in wet places, *R. betularum* Hora with birch in grassy copses and *R. sardonia* Fries with pines. Brief notes are here included giving the basic characters of eight common species, but it must be appreciated the identification of many species within this genus is difficult.

R. atropurpurea (Krombholz) Britz.   Blackish purple russula

Cap: width 50–100 mm. *Stem:* width 14–25 mm; length 60–80 mm.

Cap: deep reddish purple but becoming spotted with either cream-colour or white blotches.

Stem: white but becoming flushed greyish or stained brownish with age.

Gills: white then very pale yellow.

Flesh: white in cap and stem.

Spore-print: white.

On the ground in mixed woods and copses, particularly those containing oak.

Plate 7. Fleshy but brittle fungi: Spores whitish and borne on gills

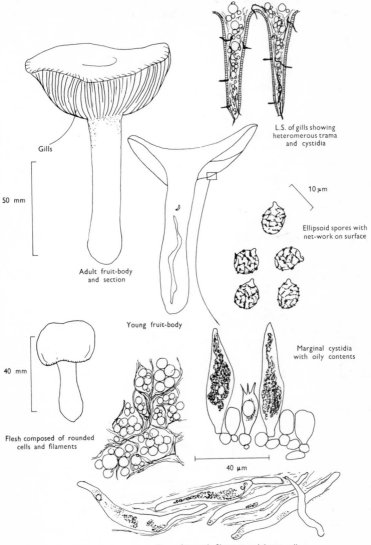

Gills

50 mm

Adult fruit-body
and section

Young fruit-body

40 mm

Flesh composed of rounded
cells and filaments

L.S. of gills showing
heteromerous trama
and cystidia

10 μm

Ellipsoid spores with
net-work on surface

Marginal cystidia
with oily contents

40 μm

Cap-surface with filaments and larger cells
filled with oily material

Russula cyanoxantha (Secretan) Fries
Cap: width 50–150 mm. *Stem:* width 10–30 mm; length 50–100 mm.
Cap: lilac, bluish to purple often with green tints.
Stem: pure white.
Gills: pure white.
Flesh: white.
Spore-print: white.
Common in deciduous woods, especially beech-woods.

R. emetica (Fries) S. F. Gray   Emetic russula
Cap: width 50–100 mm. *Stem:* width 8–15 mm; length 25–70 mm.
Cap: bright scarlet fading with age to become spotted pinkish, slightly
    viscid when moist.
Stem: spongy, fragile.
Flesh: white.
Gills: pure white.
Spore-print: pure white.
In pine woods usually in boggy areas.

R. fellea (Fries) Fries   Geranium-scented russula
Cap: width 40–75 mm. *Stem:* width 10–20 mm; length 30–75 mm.
Cap: tacky when fresh, straw-coloured or pale tawny brown.
Stem: similarly coloured to the cap.
Gills and flesh: pale straw-colour and smelling of House Geraniums
    (i.e. Pelargoniums).
Spore-print: cream-coloured.
Common under beech.

R. foetens (Fries) Fries   Foetid russula
Cap: width 70–170 mm. *Stem:* width 15–30 mm; length 50–90 mm.
Cap: slimy, dingy yellow to tawny, margin strongly furrowed and
    ornamented with raised bumps.
Stem: whitish then flushed or spotted with rust-brown.
Gills: straw-coloured, often spotted brown with age and beaded with
    watery droplets when growing under moist conditions.
Flesh: white to cream, brittle and with foetid-oily smell.
Spore-print: pale cream-colour.
Common in deciduous woods.

R. mairei Singer

Cap: width 30–75 mm. *Stem:* width 7–15 mm; length 35–70 mm.

Cap: scarlet red but developing creamy areas with age, dry.

Stem and gills: white but with a distinct although faint greenish grey flush, the former fairly firm.

Flesh: white.

Spore-print: pure white.

Commonly accompanying beech, even individual trees in gardens.

R. nigricans (Mérat) Fries   Blackening russula

Cap: width 75–200 mm. *Stem:* width 15–35 mm; length 25–75 mm.

Cap: cream-coloured then flushed sooty brown, finally black as if scorched by proximity to bonfire.

Stem: white then dark brown.

Gills: pale ochre reddening when bruised, thick and very distant.

Flesh: white slowly dull red on cutting then brown and finally changing soot-colour after some time.

Spore-print: white.

Common in deciduous woods.

R. xerampelina (Secretan) Fries

Cap: width 50–140 mm. *Stem:* width 15–30 mm; length 40–60 mm.

Cap: deep blood-red or brownish red.

Stem: white with a flush of red towards the base.

Gills: cream then ochraceous.

Flesh: white staining brownish and smelling strongly of fish- or crab-paste, and staining dark green when a crystal of green iron sulphate is rubbed into it.

Spore-print: deep cream-colour.

Common in mixed woods; a very variable fungus with many colour-forms, but easily recognised by the green reaction with ferrous sulphate.

Lactarius turpis (Weinm.) Fries   Ugly milk-cap
Cap: width 60–200 mm. *Stem:* width 10–25 mm; length 40–75 mm.
Description:

Cap: firm, convex usually with a central depression at maturity, dark olive-brown or dark greyish olive with a yellow-tawny, woolly margin when young which soon disappears, and the whole cap becomes sticky with age and turns deep purple when a drop of household ammonia is placed on it.

Stem: short, stout, similarly coloured to the cap except for the distinctly ochraceous apex, slimy and pitted.

Gills: crowded, cream-coloured to pale straw-coloured, but soon spotted with dirty brown, particularly when bruised.

Flesh: white or greyish ochre exuding a milk-like liquid which lacks a distinct smell and is white and unchanging when exposed to the air.

Spore-print: pale pinkish buff.

Spores: subglobose or ellipsoid and covered in a network of strongly developed, raised lines interconnected by finer ones, both of which stain blue-black in solutions containing iodine, generally 8×6 μm in size (7–8×6–7 μm).

Marginal cystidia: lance- or spindle-shaped and filled with oily contents.

Facial cystidia: similar to marginal cystidia.

Habitat & *Distribution:* Common in woods and copses, or on heaths especially in boggy places but always where birch is growing.

General Information: Easily recognised by the dull colours and purple reaction with alkali; there is no British species with which *L. turpis* can be mistaken. The purple reaction is similar to that found in the familiar school laboratory reagent litmus, for the compound found in *L. turpis* turns purple in alkali and reddens in acidic solutions. First discovered by Harley in 1893 this reaction marked the beginning of a whole series of chemical studies on the agarics which has led to the discovery of many unique compounds.

Illustrations: Hvass 214 (but too green); LH 213; NB 113[3]; WD 38[1].

General notes on the genus *Lactarius*

There is little doubt that the genus *Russula* and the genus *Lactarius* are closely related; in fact they stand aside from the other agarics in

Plate 8. Fleshy and milking fungi: Spores whitish and borne on gills

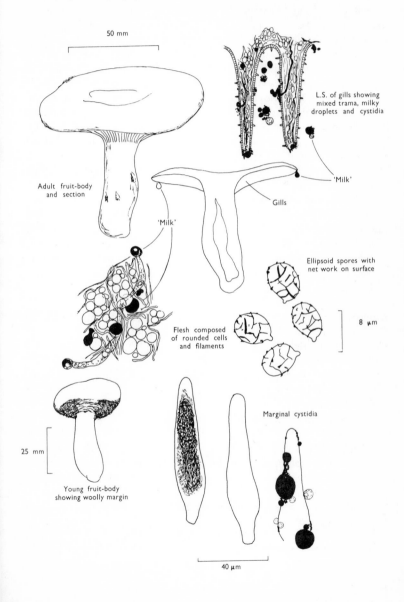

50 mm

L.S. of gills showing
mixed trama, milky
droplets and cystidia

'Milk'

Adult fruit-body
and section

Gills

'Milk'

Ellipsoid spores with
net work on surface

8 μm

Flesh composed
of rounded cells
and filaments

25 mm

Marginal cystidia

Young fruit-body
showing woolly margin

40 μm

the very important character mentioned on page 46. In Europe the easiest distinction between the two genera is that members of the genus *Lactarius* exude a milk-like juice which may be white or variously coloured depending on the species involved (e.g. purple in *L. uvidus* (Fries) Fries, yellow in *L. chrysorheus* Fries). The cap, stem and frequently the gills are brittle and when broken liberate the milk-like liquid; when the fruit-body is dry, however, the presence of this liquid may be difficult to demonstrate. The spores have a blue-black ornamentation under the microscope when mounted in iodine, and although when in mass the colours are not as varied as those found in the genus *Russula* there is every likelihood that they will play an important role in the classification of the group in the future. The colour of the spore-print has been rather neglected, although the genus includes some rather unusual fungi.

The odours of many species are very distinct and vary from the smell of coconut and spice to those of various flowers; an odour commonly met with is termed 'oily rancid resembling butter which has become mouldy'; in early books it was described as being the smell of bed-bugs!

The majority of the species are undoubtedly mycorrhizal: thus *L. torminosus* is found with birch, *L. deliciosus* and *L. rufus* with conifers and *L. quietus* with oak. Brief notes are given on additional species:—

L. camphoratus (Fries) Fries   Curry-centred milk-cap
Cap: width 20–50 mm. *Stem:* length 20–50; width 4–6 mm.
Cap and stem: red-brown.
Gills: reddish brown.
Flesh: reddish buff with an aromatic odour resembling spices which becomes very strong when dried and exudes a pale thin milk-like liquid.
Common in conifer woods and plantations.

L. deliciosus (Fries) S. F. Gray   Saffron milk-cap
Cap: width 50–120 mm. *Stem:* length 20–60 mm; width 15–25 mm.
Cap: viscid, dirty greyish ochre with flush of tawny but soon becoming greenish with age.
Stem: dirty buff or greyish ochre, spotted with green particularly with age or on handling.
Gills: orange-yellow bruising deep orange but becoming green with time.

Flesh: pinkish to apricot-coloured but becoming green with age and exuding a rich orange-red fluid which gradually becomes greyish green.

Frequent in conifer woods and plantations.

L. glyciosmus (Fries) Fries   Coconut-scented milk-cap
Cap: width 20–50 mm. *Stem:* length 30–50 mm; width 5–8 mm.
Cap: usually with a central 'bump', greyish lilac, dull and minutely scaly or velvety.
Stem: white to pale yellowish.
Gills: pale yellowish to flesh-coloured then flushed lilaceous.
Flesh: pale yellowish or flushed lilaceous, smelling strongly of desiccated coconut and exuding a white unchanging milk-like liquid.
In woods and on heaths, particularly where birch is growing.

L. quietus (Fries) Fries   Oak milk-cap
Cap: width 30–80 mm. *Stem:* length 40–80 mm; width 10–15 mm.
Cap and stem: milky cocoa-coloured, zoned with reddish brown.
Gills: pale ochraceous then flushed red-brown.
Flesh: similar to gills, smelling strongly of rancid oil, and exuding a white, thin milk-like liquid which becomes very, very faintly yellow on exposure to the air.
Common wherever oak is growing.

L. rufus (Fries) Fries   Rufous milk-cap
Cap: width 50–90 mm. *Stem:* length 50–90 mm; width 10–15 mm.
Cap: dark red-brown with a distinct, usually sharp 'bump' in centre.
Stem: pale red-brown throughout or whitish at base.
Gills: pale reddish brown and exuding a white, unchanging milk-like fluid.
In pine woods and less frequently with birches on acid heaths.

L. torminosus (Fries) S. F. Gray   Woolly milk-cap
Cap: width 40–150 mm. *Stem:* length 60–100 mm; width 15–30 mm.
Cap: pale strawberry-pink or pale salmon colour, distinctly zoned, slimy when wet at centre and strongly shaggy fibrillose at margin.
Stem and gills: pale strawberry colour.
Flesh: tinged salmon-pink and exuding a white unchangeable milk-like liquid.
Frequent where birches grow.

Amanita muscaria (Fries) Hooker   Fly agaric
Cap: width 100–175 mm. *Stem:* width 30–40 mm; length 150–275 mm.
Description:
Cap: bright scarlet to orange-red with scattered whitish or yellowish
    fragments of veil particularly towards the centre and hanging down
    from the margin, viscid when moist, striate at margin with age.
Stem: white, striate above the soft easily torn, although prominent,
    ring which is white above and yellow below; stem-base swollen and
    ornamented with patches of yellowish or white veil-fragments which
    form concentric rings or ridges of tissue.
Gills: white, free, crowded, fairly thick, minutely toothed at their edge.
Flesh: soft, lacking distinctive smell, or at times slightly earthy and
    white, yellowish below cap-centre.
Spore-print: white.
Spores: long, hyaline under the microscope, ellipsoid, smooth about
    10 × 7 μm in size (10–13 × 7–8 μm).
Marginal cystidia: composed of chains of swollen, hyaline cells.
Facial cystidia: absent.
Habitat & *Distribution:* Found in birch-woods, less frequently collected
    in the vicinity of conifers; wide-spread and fairly common, but it is
    erratic in its appearance giving the impression of being absent from
    a locality until one season it suddenly fruits in profusion.
General Information: An easily recognised fungus because of its striking
    colour. It is also very familiar and well-known because it appears so
    often on Christmas cards, and features commonly in illustrations in
    children's story-books. The fungus contains a poison which formerly
    was used to kill flies—hence the common name of 'Fly agaric' and
    the scientific name from the latin name for the house-fly. The red
    skin of the cap, where the major amount of the poison resides, was
    cut up with a little milk and sugar or honey; flies attracted to this
    sweet concoction inadvertently ate the poison and later perished.
    This fungus has a very well documented and long history and appears
    in the legends of many countries. It is featured in Greek mythology,
    Slavic and Scandinavian folk-lore and indeed appears in the pre-
    history of Indian tribes of N.E. Asia. It has even been connected
    with the formation of certain sects within the early Christian church.
Illustrations: F. frontispiece; Hvass 1; LH 117; NB 113[1]; WD 2[1].

Plate 9. Fleshy fungi: Spores white and borne on gills

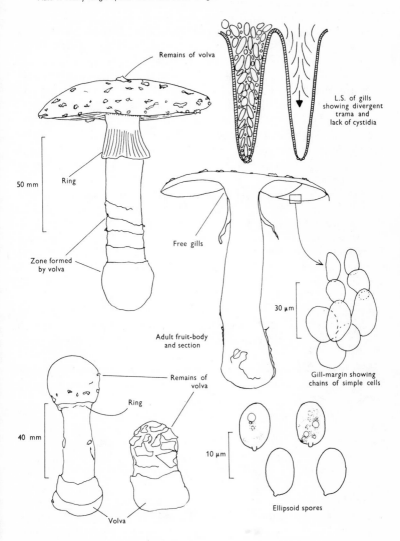

Remains of volva

L.S. of gills showing divergent trama and lack of cystidia

Ring

50 mm

Zone formed by volva

Free gills

30 µm

Adult fruit-body and section

Gill-margin showing chains of simple cells

Remains of volva

Ring

40 mm

10 µm

Volva

Ellipsoid spores

Young fruit-bodies showing the elongation of stem and disruption of volva

Notes on the genus *Amanita*

The genus *Amanita* contains many important mycorrhizal fungi including the 'Blusher', *A. rubescens* (Fries) S. F. Gray, the 'Tawny grisette', *A. fulva* Secretan, and the 'False death-cap', *A. citrina* S. F. Gray. The first grows on heaths and in woods with a variety of trees; *A. fulva* frequently grows with birch and *A. citrina* with several leafy trees although its var. *alba* (Gillet) E. J. Gilbert appears to be confined to beech woods. However, there is some evidence that many members of the genus in drier more southern countries than Britain, are non-mycorrhizal. In fact the genus as a whole may be southern-temperate in its distribution. In the British Isles the number of species of *Amanita* recorded decreases as one goes north, or the frequency of single species except for a few widespread forms falls off northwards. In a few cases a more familiar southern species is replaced in similar habitats by another species, e.g. *A. phalloides* (Fries) Secretan is replaced by *A. virosa* Secretan the 'Destroying angel' in Scotland, and *A. citrina* frequently in the north by *A. porphyria* (Fries) Secretan. Species of *Amanita* are usually large conspicuous fungi and the genus contains some of our best known agarics. One, *A. muscaria* (Fries) Hooker has already been mentioned, but the genus also includes the 'Death-cap' *A. phalloides* and 'Caesar's mushroom' *A. caesarea* (Fries) Schweinitz, a fungus not found in this country but considered to be superior in edibility to all other fungi; thus edible and deadly poisonous species are found closely related and this simply emphasises how important it is not to eat the agarics one finds in the woods and fields except when accompanied by a 'real' expert. Deaths or near fatalities in Europe and North America are recorded annually due to the eating of fungi belonging to this genus.

The poisonous qualities of the fungi in this genus—only a very small amount of poison is often sufficient to produce fatal results—has led to a close connection between these fungi and black magic and the supernatural. This connection is even more emphasised when it is learnt that some have an intoxicating effect. Hence the long history mentioned earlier.

Members of the genus *Amanita* are characterised by their anatomy and certain macroscopic features; the former is illustrated under *A. muscaria*, i.e. the divergent gill-trama. The main macroscopic character of note is the presence of a volva at the base of the stem and it is the details of this volva which helps to distinguish different species. *A. phalloides* has a distinct, loose, membranous sheath, in *A. citrina* the volva is reduced to a narrow rim around the bulbous stem and in *A. rubescens* and *A. muscaria* the volva is simply a series of concentric zones of woolly scales. All the four species noted above possess a ring, but *A. fulva* the 'Tawny grisette' and *A. vaginata* (Fries) Vittadini the 'Grisette' only possess a volva; this has lead to the use of the generic name *Amanitopsis* in many books, now no longer considered necessary.

The veil in *Amanita* is probably the most highly developed amongst our common agarics and from Appendix iv it can be seen how the scaly cap and stem originate and how the volva differs from the ring. The volva and cap-scales constitute what has been called the universal veil and the ring which stretches from the cap margin to the stem has been termed the partial veil.

The spores of species of *Amanita* are large and their shape and chemical reactions help to distinguish the different species within the genus. One of the most interesting features, however, is that the spore-mass, although usually described as white, in many species is not white but flushed greenish grey, etc. The slight subtleties in colour of the spore-print assist in classification.

The following notes may be instructive in conjunction with the information above (for common names see above).

(i) Possessing a ring on the stem:—

A. citrina S. F. Gray
Cap: width 55–80 mm. *Stem:* width 18–22 mm; length 70–80 mm.
A lemon-yellow or whitish capped agaric with bulbous stem-base, white patches of volva on cap and white stem with flesh strongly smelling of new potatoes.
Spores: almost globose and measuring 9–10 × 7–8 μm.

A. excelsa (Fries) Kummer
Cap: width 75–140 mm. *Stem:* width 20–28 mm; length 85–120 mm.
A greyish or brownish capped agaric with clavate stem-base, grey patches of volva on the cap and white concentrically scaly stem with flesh unchanged on exposure to the air.
Spores: broadly ellipsoid and measuring 9–10 × 8–9 μm.

A. rubescens (Fries) S. F. Gray

Cap: width 70–120 mm. *Stem:* width 12–25 mm; length 65–100 mm.

A reddish fawn or pinkish buff capped agaric with swollen stem-base, pinkish or flesh-coloured patches of volva on cap and reddish concentrically scaly stem with flesh becoming reddish when exposed to the air.

Spores: ellipsoid and measuring 9–10 $\times$ 5–6 μm.

A. pantherina (Fries) Secretan   'Panther'

Cap: width 48–95 mm. *Stem:* width 12–20 mm; length 65–100 mm.

An olive-brown or smoky brown capped agaric with only slightly swollen stem-base, white patches of volva on the cap and white concentrically scaly stem with unchanging flesh.

Spores: ellipsoid and measuring 8–12 $\times$ 7 μm.

A. phalloides (Fries) Secretan

Cap: width 70–85 mm. *Stem:* width 12–20 mm; length 85–120 mm.

A greenish or yellow-olive capped agaric with stem sheathed in membranous volva, white patches of volva on cap and smooth, white stem with white flesh.

Spores: broadly ellipsoid and measuring 10–12 $\times$ 7 μm.

(ii) Lacking ring on stem:—

A. fulva Secretan

Cap: width 40–60 mm. *Stem:* width 10–15 mm; length 100–150 mm.

A thin, tawny-brown agaric with stem sheathed in membranous volva and pale tawny, slightly scaly stem.

Spores: globose and 10–12 μm in diameter.

A. vaginata (Fries) Vittadini

Differs from *A. fulva* in the cap being metallic grey or silvery in colour.

(b) Parasites

Armillaria mellea (Fries) Kummer   Honey-fungus
Cap: width 50–150 mm. *Stem:* width 10–12 mm; length 75–150 mm.
Description: Plate 10.

Cap: at first convex then more or less flattened or slightly depressed, very variable in colour, yellowish, olive, buff, sand-coloured or some shade of brown, at first covered in small, brownish or ochraceous scales which give the young cap a velvety aspect, but gradually the scales disappear with age except at the cap-centre; margin striate and usually paler than centre of the cap.

Stem: equal or swollen at base, often several grouped together, white at apex above a whitish, rather thick, ring which is flushed with olive-yellow or red-brown at its margin; stem-base fibrillose, whitish but finally red-brown at maturity.

Gills: adnate or slightly decurrent, whitish then flushed flesh colour and developing brownish spots with age or in cold, wet weather.

Flesh: with rather strong and unpleasant smell, white or flushed pinkish in the cap, brown and stringy in the stem.

Spore-print: very pale cream colour.

Spores: medium-sized, hyaline, ellipsoid, less than 10 µm in length (8–9 × 5–6 µm).

Marginal cystidia: variable, hyaline, cylindric and not well-differentiated.

Facial cystidia: absent.

Habitat & Distribution: This fungus grows in troops or is found joined at the base to form clusters. It is always attached to old trees, trunks, stumps and buried wood, either directly or by its vegetative stage which darkens and aggregates to form strands resembling boot-laces which are called rhizomorphs.

General Information: This rather variable, and therefore often perplexing, fungus causes a destructive rot of trees and can travel long distances through the soil with the use of its rhizomorphs. It commonly grows on several species of broad-leaved trees, but can also colonise conifer trees. It also attacks garden shrubs, such as privet-hedges, and is particularly destructive to Rhododendrons causing a wilt of the whole shrub and subsequent death; it has also been recorded as attacking potatoes. The actively growing mycelium which can often be found growing under the bark of infected trees, exhibits a lumin-

osity if freshly exposed and placed in a darkened room. The rhizo-morphs of *A. mellea* are highly specialised structures composed of mycelial threads some of which have become rather more differen-tiated than is normally found in the vegetative stage of other agarics.
Illustrations: F 27a; Hvass 26; LH 93; NB 141[1]; WD 4[3].

Pholiota squarrosa (Fries) Kummer   Shaggy Pholiota
Cap: width 50–120 mm. *Stem:* width 17–25 mm; length 95–125 mm.
Description: Plate 11.
Cap: convex, but expanding and becoming flattened with a slight central umbo, ochre-yellow to yellowish rust-colour and covered with dark brown recurved scales which are particularly dense at the centre.
Stem: variable in length and thickness depending on how it is attached to the substrate, whether in a deep crack or wound, or in a depression, and how many specimens are in the cluster; its colour is similar to that of the cap, exhibits a small, dark brown fibrillose, torn ring or ring-zone and is ornamented with recurved red-brown scales below that ring.
Gills: broadly adnate with a decurrent tooth and crowded, yellowish at first then rust-coloured.
Flesh: with strong, pleasant but pungent smell, yellowish brown, soft in the cap, fibrous in the stem.
Spore-print: rich rust-brown.
Spores: medium-sized, pale brown under the microscope, smooth, ellipsoid, and 6–8 × 4 μm in size.
Marginal cystidia: spindle-shaped, hyaline, numerous.
Facial cystidia: flask-shaped with a small apical appendage and be-coming rich yellow when immersed in solutions containing ammonia.
Habitat & *Distribution:* Common in clusters in woods, gardens or parks, on wood or at the base of the trunks of broad-leaved trees in summer and autumn.

Plate 10. Fleshy fungi: Spores white and borne on gills

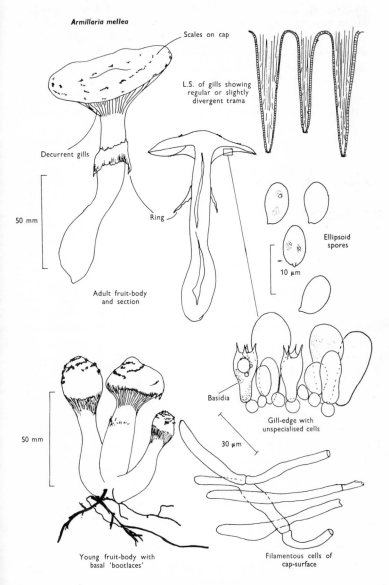

Armillaria mellea

Scales on cap

Decurrent gills

Ring

50 mm

Adult fruit-body and section

L.S. of gills showing regular or slightly divergent trama

Ellipsoid spores

10 µm

Basidia

Gill-edge with unspecialised cells

30 µm

50 mm

Young fruit-body with basal 'bootlaces'

Filamentous cells of cap-surface

General Information: Although rather a common easily recognisable and aesthetically pleasing fungus growing in its characteristic clusters at the base of trees, it is a weak parasite entering the living tissue after invading decayed areas of the tree. This is the reason why when branches are broken off trees by wind, snow or storms, they should be carefully trimmed to remove ragged edges and the wound treated with a protective tar to stop the entry of rain, cold and fungus spores. Other more destructive fungi may enter a tree through such wounds; *P. squarrosa* frequently attacks mountain ash or rowan.

It is recognised by the dry scaly cap and stem which helps to distinguish it from the sticky capped *P. aurivella* (Fries) Kummer with similar habitat preferences but wider spores (6–9 × 4–5 μm). *P. adiposa* (Fries) Kummer is found on beech trees and it, too, has a viscid cap, but the spores are 5–6 × 3–4 μm in dimensions.

Illustrations: Hvass 134; LH 149; WD 54[2].

Plate 11. Fleshy fungi: Spores rust-brown and borne on gills

Pholiota squarrosa

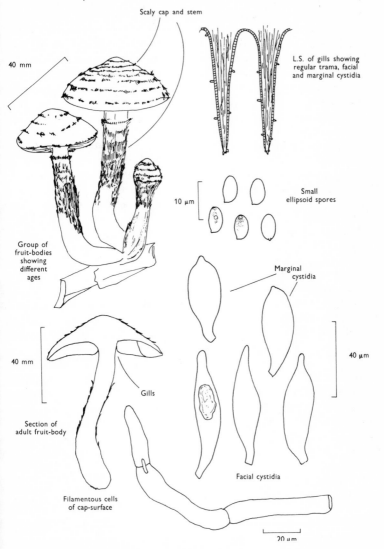

Scaly cap and stem

L.S. of gills showing regular trama, facial and marginal cystidia

40 mm

10 µm

Small ellipsoid spores

Group of fruit-bodies showing different ages

Marginal cystidia

40 mm

40 µm

Gills

Section of adult fruit-body

Facial cystidia

Filamentous cells of cap-surface

20 µm

(c) Saprophytes—wood inhabiting or lignicolous agarics

Hypholoma fasciculare (Fries) Kummer   Sulphur-tuft
Cap: width 20–50 mm. *Stem:* width 6–13 mm; length 40–100 mm.
Description:
Cap: sulphur-yellow, flushed with sand-colour or red-brown at centre
    then ochraceous yellow throughout, convex at first with margin
    incurved and clothed with fibrillose remnants of a yellow-olive veil,
    but then becoming flattened and losing evidence of that veil.
Stem: equal or flexuous, usually with several joined at base, similarly
    coloured to the cap, fibrillose streaky or with some fibrils from the
    veil stretching from the cap to the stem in young specimens.
Gills: sinuate and crowded, at first sulphur-yellow then olive-green,
    but finally with a flush of purple-brown.
Flesh: with rather strong and unpleasant smell, yellow throughout.
Spore-print: purple-brown.
Spores: medium-sized, ellipsoid or ovoid, smooth, purple-brown and
    less than 10 μm in length (6–8 × 4 μm).
Marginal cystidia: flasked-shaped, short, cylindric and hyaline.
Facial cystidia: more swollen than marginal cystidia and with silvery
    contents which yellow in solutions containing ammonia.
Habitat & *Distribution:* The sulphur-tuft grows in dense clusters on
    and around old stumps of broad-leaved trees, and can be found
    throughout the year; it also grows on conifers, but less frequently.
General Information: It may be recognised by the greenish tint of the
    immature gills and of the young cap. *H. capnoides* (Fries) Kummer
    grows on the wood of coniferous trees and has a much more ochrace-
    ous brown cap and stem than the sulphur-tuft and slightly larger
    spores—7–8 × 4–5 μm. *H. sublateritium* (Fries) Quélet grows on
    hardwoods but is bigger than *H. fasciculare* and has a brick-coloured
    cap and very sturdy stem (spores 6–7 × 3–4 μm).
Illustrations: F 37b; Hvass 176; LH 147; NB 141[5]; WD 76[2].

Plate 12. Fleshy fungi: Spores purplish brown and borne on gills

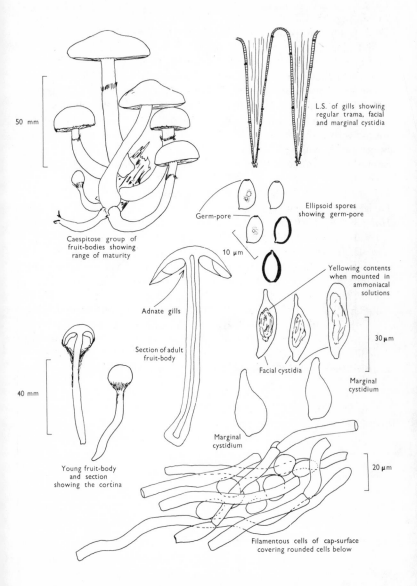

50 mm

L.S. of gills showing regular trama, facial and marginal cystidia

Germ-pore

Ellipsoid spores showing germ-pore

Caespitose group of fruit-bodies showing range of maturity

10 μm

Adnate gills

Section of adult fruit-body

Yellowing contents when mounted in ammoniacal solutions

30 μm

Facial cystidia

Marginal cystidium

40 mm

Marginal cystidium

Young fruit-body and section showing the cortina

20 μm

Filamentous cells of cap-surface covering rounded cells below

Flammulina velutipes (Fries) Karsten   Velvet-shank
Cap: width 20–80 mm. *Stem:* width 5–10 mm; length 35–60 mm.
Description:

Cap: bright sand-colour or slightly red-brown at centre, convex at
   first then flattened with age, smooth, slimy because of the presence
   of a sticky elastic skin, rather rubbery to the touch.

Stem: cylindrical or slightly swollen towards the base, dark brown and
   densely hairy or velvety, tough and rubbery to handle.

Gills: adnexed, very unequal and somewhat distant, pale yellow, gradually becoming buff as the spores mature.

Flesh: with rather pleasant smell, yellowish, watery and soft.

Spore-print: white.

Spores: medium-sized, hyaline, ellipsoid and about 8 × 3–4 μm in
   size (7–9 × 3–4 μm).

Marginal cystidia: hyaline, elongate, broadly flask-shaped.

Facial cystidia: similar to marginal cystidia.

Habitat & Distribution: Found in clusters on old stumps, fallen trunks
   and on the wounded parts of standing trees.

General Information: This fungus can be recognised by the clustered
   habit, the viscid, bright tawny cap and the dark velvety stem. This
   is one of the few agarics which occurs regularly late in the season,
   even appearing in the winter, although it can be seen growing in
   its familiar groups at almost any time of the year. This fungus holds
   a rather isolated position in classification and was once placed in the
   genus *Collybia*. It may be found in several books under this last
   genus.

Illustrations: F 18b; Hvass 80; LH 109; NB 141[3]; WD 21[4].

Plate 13. Fleshy fungi: Spores white and borne on gills

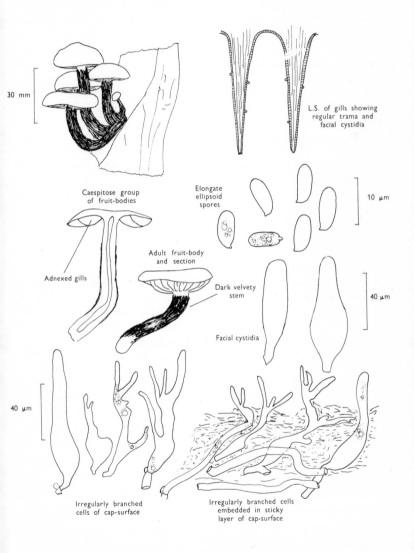

30 mm

Caespitose group
of fruit-bodies

L.S. of gills showing
regular trama and
facial cystidia

Elongate
ellipsoid
spores

10 μm

Adnexed gills

Adult fruit-body
and section

Dark velvety
stem

40 μm

Facial cystidia

40 μm

Irregularly branched
cells of cap-surface

Irregularly branched cells
embedded in sticky
layer of cap-surface

Mycena galericulata (Fries) S. F. Gray   Bonnet mycena
Cap: width 25–50 mm. *Stem:* width 3–6 mm; length 50–125 mm.
Description:
Cap: conical or bell-shaped then expanding but retaining a central
    umbo, never completely flattened, smooth, greyish, pale sepia or
    dirty white and striate with darker lines from the margin to the
    centre.
Stem: similarly coloured to the cap, smooth, shiny, tough and usually
    noticeably downy at base.
Gills: at first white flushed distinctly pale pink with age, uncinate,
    rather distant and sometimes with interconnecting veins.
Flesh: white with little or no distinctive smell.
Spore-print: white.
Spores: medium-sized, hyaline, broadly ellipsoid, smooth, about 10 ×
    7 μm in size (9–12 × 6–8 μm) and staining bluish grey when mounted
    in solutions containing iodine.
Marginal cystidia: club-shaped but the apex ornamented with blunt
    hairs of varying lengths.
Facial cystidia: absent.
Habitat & *Distribution:* Commonly found, in all but the coldest months,
    in woods, parks or gardens, often in dense clusters on stumps and
    fallen trunks of broad-leaved trees.
General Information: This is one of our commonest members, and one
    of the largest in the genus *Mycena*; many species in this genus are
    quite small yet are nevertheless very important components of the
    woodland flora decomposing leaves, twigs, etc., and contributing in
    this way to the recirculating of organic matter.
    The name *Mycena* is derived from the same Greek word as that
which refers to the country around the ancient city of Mycenae in
the plain of Argos, and from whence Agamemnon came and gathered
his forces to invade Troy to reclaim Helen his wife. It has been
suggested that this similarity in name came about through the
necessity for an army stationed in Argos, early in the history of
Ancient Greece, to rely on the mushrooms found on the plains
about to save the soldiers from starvation.
Illustrations: F 17a; Hvass 119; LH 109; NB 133[8]; WD 26[3].

Plate 14. Fleshy fungi: Spores white and borne on gills

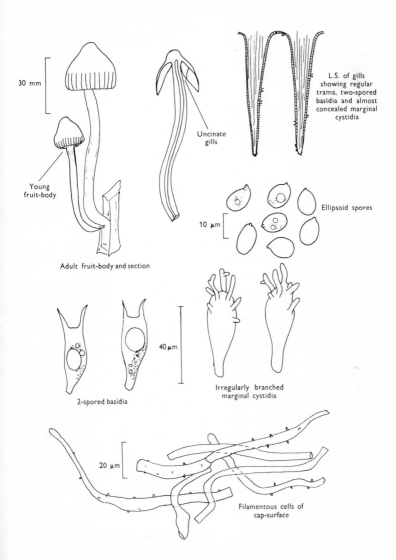

30 mm

Young
fruit-body

Adult fruit-body and section

Uncinate
gills

L.S. of gills
showing regular
trama, two-spored
basidia and almost
concealed marginal
cystidia

10 μm

Ellipsoid spores

40 μm

2-spored basidia

Irregularly branched
marginal cystidia

20 μm

Filamentous cells of
cap-surface

Pluteus cervinus (Fries) Kummer   Fawn pluteus
Cap: width 40–100 mm. *Stem:* width 10–15 mm; length 75–125 mm.
Description:

Cap: conical, rapidly expanding and then becoming plano-convex or
  flattened with only a slight but persistent umbo, dark brown, umber
  or vandyke brown, viscid when wet and often with radiating fibrils.

Stem: white, streaked to varying degrees with dark brown fibrils,
  cylindrical or slightly swollen towards the base, where it is attached
  to the substrate.

Gills: remote, very crowded, thin, at first white then distinctly salmon-
  pink.

Flesh: with pleasant smell, white and soft.

Spore-print: dull salmon-pink.

Spores: medium-sized, very faintly buff under the microscope,
  broadly ellipsoid and 7–8 × 5–6 μm in size.

Marginal cystidia: flask-shaped, the majority with three or four hooks
  at the distinctively thick-walled apex.

Facial cystidia: similar to marginal cystidia but sometimes intermixed
  with those lacking hooks.

Habitat & Distribution: This fungus grows singly or in groups on old
  stumps and fallen trunks throughout the year except for the most
  wintry months; it is commonest in autumn.

General Information: This fungus may also grow on old sawdust heaps,
  a habitat which is often very worth while examining in detail by the
  interested amateur during wet seasons. In summer sawdust heaps
  dry out but after a good soaking, which, of course, can be applied
  artificially by frequent watering with a hose or watering-can, many
  interesting fungi develop. On sawdust heaps containing conifer
  debris a larger species with black or dark brown edge to the gills is
  found—*P. atromarginatus* Kühner.

  The peculiar pointed cystidia found on the gill-edge and on the
  gill-face of *P. cervinus* were thought by some early mycologists to
  stop mites and insect larvae from crawling up between the gills and
  damaging the developing spores. There is no evidence that this
  actually takes place in nature; the real purpose of these obscure
  structures is unknown and has been little studied.

Illustrations: Hvass 127; LH 121; NB 135[1]; WD 50[2].

Plate 15. Fleshy fungi: Spores pinkish and borne on gills

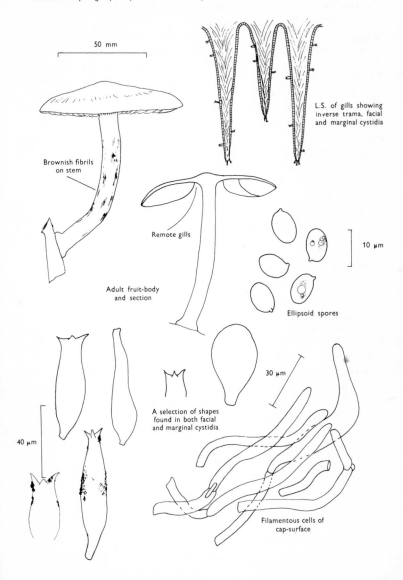

50 mm

L.S. of gills showing inverse trama, facial and marginal cystidia

Brownish fibrils on stem

Remote gills

10 μm

Ellipsoid spores

Adult fruit-body and section

30 μm

A selection of shapes found in both facial and marginal cystidia

40 μm

Filamentous cells of cap-surface

Gymnopilus penetrans (Fries) Murrill

Cap: width 20–50 mm. *Stem:* width 4–7 mm; length 20–50 mm.
Description:

Cap: convex then becoming flattened at maturity, dry, slightly scaly, golden tawny, or rusty yellow and when young with the remnants of a rapidly disappearing yellow cortina hanging from the margin.

Stem: yellow above and red-brown or orange-tawny below and darkening on bruising; veil forming a delicate fibrillose zone in the upper part of the stem which is soon lost on excessive handling.

Gills: adnate to slightly decurrent, thin and crowded, at first golden yellow, but soon spotted rust colour.

Flesh: yellow and lacking distinctive smell.

Spore-print: rich orange-tawny.

Spores: medium-sized, ellipsoid, finely roughened and deep yellow brown under the microscope, less then 10 μm in length (7–8 × 5–4 μm).

Marginal cystidia: hyaline, flask-shaped with long often slightly irregular neck.

Facial cystidia: similar to the marginal cystidia, but often broader.

Habitat & *Distribution:* This fungus is found on sticks or twigs or chips of coniferous wood, particularly in plantations.

General Information: Although it has only comparatively recently been recognised in Britain it is very wide-spread. It has been confused with, indeed described under, the name of the less-common fungus *Gymnopilus sapineus* (Fries) Maire which also grows in conifer woods; it is easily distinguished, however, by its spotted gills. Both the fungi above can be found in books under the old name *Flammula*, from the bright colour of the caps of many of its constituent members, but *Flammula* has been used for a genus of flowering plants also and this has precedence.

Illustrations: F 29a; Hvass 152 not very good; LH 175 not very good; NB 109[6].

Plate 16. Fleshy fungi: Spores rust-brown and borne on gills

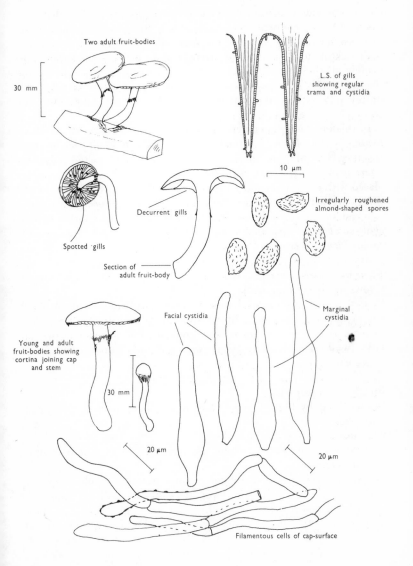

Two adult fruit-bodies

30 mm

L.S. of gills showing regular trama and cystidia

10 μm

Decurrent gills

Irregularly roughened almond-shaped spores

Spotted gills

Section of adult fruit-body

Young and adult fruit-bodies showing cortina joining cap and stem

Facial cystidia

Marginal cystidia

30 mm

20 μm

20 μm

Filamentous cells of cap-surface

Notes on the artificial family group '*Pleurotaceae*'—the Oyster mushrooms

One of the common features of lignicolous fungi is the fact that they lack a distinct stem or if one is present it is attached to one side of the cap, i.e. lateral. However, in the past the correlation of the habitat with lack of stem has induced mycologists to define a single family to include all these forms. After studying the anatomy and microscopic characters this grouping has been found to be entirely artificial and simply reflects how the morphology is tied up intimately with the ecology of a species.

In this one family members of the genera *Panus*, *Panellus*, *Lentinus*, *Lentinellus*, *Crepidotus*, *Pleurotellus*, and *Pleurotus* have all been grouped together, but some of the genera are more related to the polypores referred to later (p. 135); many of those with brown spores are better placed with *Cortinarius* and some of those with white or cream-coloured spores are better placed close to *Mycena* and *Tricholoma*. This leaves as a residue the genus *Pleurotus*, a genus which although rather heterogeneous contains one familiar member, i.e. the common Oyster mushroom, *Pleurotus ostreatus*.

Pleurotus ostreatus (Fries) Kummer   Oyster mushroom
Grows up to 150 mm across.
Cap: flattened, shell-shaped, smooth or slightly cracked, deep bluish grey, gradually becoming brownish with age and finally dark buff.
Stem: absent or very short, passing gradually into one side of the cap.
Gills: white flushing dirty yellow with age, rather distant and deeply decurrent.
Flesh: white, soft and with very pleasant smell.
Spore-print: pale lilac.
Spores: long, hyaline, oblong under the microscope and 10–11 × 4 μm in size.
Marginal and facial cystidia: absent.
Habitat & Distribution: Common, clustered in tiers on stumps, trunks, posts, etc.
General Information: This fungus is not infrequent on old telephone-poles and forms white sheets of mycelium immediately under the bark of fallen trees. Although frequent in autumn it may be found throughout the year and is easily recognised by its size and bracket-

Plate 17. Wood-inhabiting, fleshy but leathery fungi: Spores whitish or brownish and borne on gills—'Pleurotaceae'

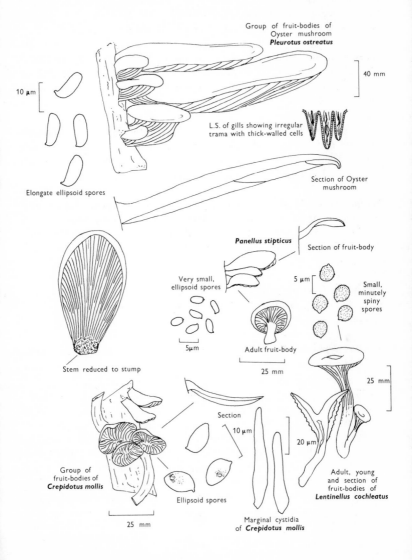

Group of fruit-bodies of Oyster mushroom
Pleurotus ostreatus

40 mm

10 μm

L.S. of gills showing irregular trama with thick-walled cells

Elongate ellipsoid spores

Section of Oyster mushroom

Stem reduced to stump

Panellus stipticus

Section of fruit-body

Very small, ellipsoid spores

5 μm

Small, minutely spiny spores

5μm

Adult fruit-body

25 mm

25 mm

Section

10 μm

20 μm

Group of fruit-bodies of **Crepidotus mollis**

25 mm

Ellipsoid spores

Marginal cystidia of **Crepidotus mollis**

Adult, young and section of fruit-bodies of **Lentinellus cochleatus**

like, shell-shaped caps. It surprisingly has a pale lilac spore-print and not as might be expected a white spore-print. In the var. *columbinus* Quélet the young caps are a beautiful peacock-blue; this variety frequently grows on poplars.

Illustrations: F 125[2]; Hvass 109; LH 107; NB 125[2]; WD 31[1].

Panus torulosus (Fries) Fries is a tough, funnel-shaped, yellowish cinnamon fungus with oblong-ellipsoid, small, hyaline spores measuring 5–6 × 3 μm and changing yellowish not bluish grey in iodine solutions.

Panellus stipticus (Fries) Karsten forms tiers of pale cinnamon-brown, more or less kidney-shaped, scurfy caps on old wood and has egg-shaped, hyaline, small spores measuring 4 × 2–3 μm which become bluish grey in iodine solutions.

Lentinellus cochleatus (Fries) Karsten forms irregular lobed and twisted, flattened or funnel-shaped dirty brownish caps with a fragrant smell, toothed gill-edges and almost spherical, small, hyaline spores measuring 5 × 4 μm which become bluish grey in iodine solutions.

Lentinellus apparently has very close affinites to *Auriscalpium*, 'the Ear pick fungus', (p. 158) both in the structure of the spores and the anatomy of the fruit-body.

Lentinus lepideus (Fries) Fries forms very tough fruit-bodies with convex or flattened, pale yellowish caps and stems ornamented with dark tawny or brown scales. The stem is often eccentric and buried in cracks or soft rotten wood on which it grows; the spores are non-amyloid. It grows on pine stumps but also on decaying or unprotected railway sleepers and wooden paving blocks, joists, etc., made of conifer wood. When the fungus fruits in a darkened environment, such as a cellar, the mushroom-like fruit-bodies are not produced but are replaced by slender branched structures similar to the 'Stag's horn' or 'Candle-snuff fungus' (p. 206), or to certain of the Fairy Club fungi (p. 172). Similar growths have been recorded for *Polyporus squamosus* which grows on hard wood timber and is described in detail later (p. 140).

Crepidotus mollis (Fries) Kummer   Soft slipper toadstool

Cap: up to 45 mm across and in tiers, sessile, shell-shaped or kidney-shaped, smooth, rubbery and brownish ochre in colour.

Gills: pale buff then cinnamon-brown and finally flushed snuff-brown, thin and crowded.

Flesh: watery, gelatinous beneath the skin of the cap and whitish buff.

Spore-print: warm brown.

Spores: ellipsoid, smooth, medium-sized, pale buff under the microscope and 8–9 × 5–5·5 μm in size.

Easily recognised by the soft elastic cap which can be stretched without breaking, the brown gills and pale buff spores. (See Plate 49, p. 153.)

Illustrations: LH 177; NB 145[3]; WD 69[1].

The artificiality of classifying all those agarics with both a spoon-shaped or bracket-shaped fruit-body, and a reduced (or lacking) stem is further exemplified by the presence of similar genera in other groups of fungi. For instance *Claudopus* is typified by pink, angular spores (Plate 28) and *Clitopilus* is characterised by longitudinally ridged spores, i.e. they are not angular in all optical sections but only when seen end on (see p. 101). An example of the former is *C. parasiticus* (Quélet) Ricken which grows on dead remains of woody fungi, and of the latter *C. passackerianus* (Pilát) Singer which may invade mushroom beds. Both species are quite small though the last fungus is similarly coloured to the more familiar *Clitopilus prunulus* (Fries) Kummer, 'The Miller', so common in woods and fields.

Thus in the British Isles agarics with eccentric stems may be found, in the white, brown and pink-spored groups—and in the tropics and subtropics the picture is completed by the existence of the genus *Melanotus* in the black-spored agarics. *M. bambusinus* Pat. grows on bamboos and *M. musae* (Berk. & Curt.) Singer grows on dead leaves and debris of bananas; the latter is also a probable agent in the decay of fibres in the tropics.

(d) Saprophytes—terrestrial agarics

Melanoleuca melaleuca (Fries) Murrill
Cap: width 40–110 mm. *Stem:* width 50–80 mm; length 50–90 mm.
Description:
Cap: dark brown, umber or vandyke when moist, hygrophanous and
  becoming very much paler on drying almost tan, convex then
  flattened sometimes umbonate, smooth or wrinkled.
Stem: white or whitish covered in brownish fibrils which increase in
  number with age or after handling; solid, rather elastic and slightly
  swollen towards the base.
Gills: white, broad, crowded and as if cut out from behind before
  joining the stem.
Flesh: with pleasant smell, soft, white, becoming brownish with age,
  particularly in the stem.
Spore-print: very pale ivory-colour.
Spores: medium-sized, ellipsoid, hyaline under the microscope and
  roughened by distinct dots which become blue-black when mounted
  in solutions containing iodine, 8×4–5 μm.
Marginal cystidia: spear- or sword-shaped, roughened with crystals
  at the top and appearing as if barbed like fish-spines.
Facial cystidia: numerous and similar to marginal cystidia.
Habitat & Distribution: Common in autumn in woods; also found in
  pastures.
General Information: A very common fungus which is rather confusing
  to the beginner because of its variation in colour, brought about by
  the change in colour with change in content of water. However, this
  fungus can be easily recognised by the unusually ornamented cystidia
  found on the gill-faces and gill-margins. This character and the
  fact that the spores possess amyloid ornamentation define in part
  the genus *Melanoleuca*. In many books this common fungus is found
  under the genus *Tricholoma*; however, members of this latter genus
  have neither amyloid ornamented spores nor barbed cystidia.
Illustrations: LH 103; WD 13[1].

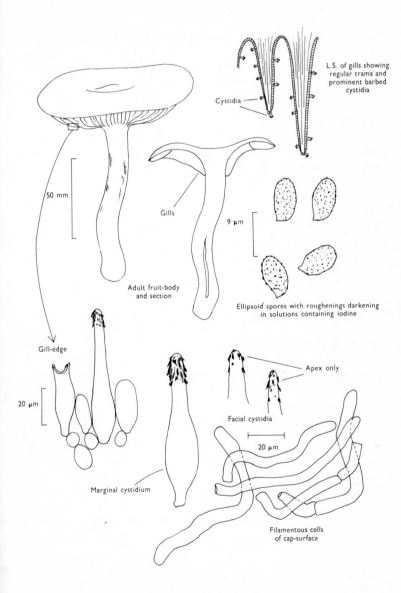

Plate 18. Fleshy fungi: Spores white and borne on gills

L.S. of gills showing regular trama and prominent barbed cystidia

Cystidia

50 mm

Gills

9 μm

Adult fruit-body and section

Ellipsoid spores with roughenings darkening in solutions containing iodine

Gill-edge

20 μm

Apex only

Facial cystidia

Marginal cystidium

20 μm

Filamentous cells of cap-surface

Clitocybe infundibuliformis (Weinm.) Quélet   Common funnel-cap
Cap: width 20–60 mm. *Stem:* width 8–13 mm; length 35–75 mm.
Description:

Cap: yellowish ochre flushed slightly pinkish buff or cinnamon but later pale tan on ageing or drying, funnel shaped.

Stem: colour like cap or slightly darker, flexible but firm and solid.

Gills: white or faintly flushed buff, decurrent and crowded.

Flesh: with pleasant slightly floral smell, white, soft and fairly thin.

Spore-print: white.

Spores: medium-sized, hyaline, tear-drop shaped, smooth, 6–7 × 3–4 μm and not blueing when mounted in solutions containing iodine.

Marginal cystidia: little different from young basidia in dimension and shape, although some may have a short apical prolongation.

Facial cystidia: absent.

Habitat & *Distribution:* Woods, copses, heaths and hill-pastures from summer to autumn.

General Information: An easily recognisable fungus because of its graceful stature, thin, funnel-shaped pinkish buff cap and tear-drop-shaped spores. Several *Clitocybe* species grow in woodlands, many of them appearing later in the season when colourful agarics are rarer.

The genus *Clitocybe* is characterised by the fleshy cap with in-curved margin when young, fibrous, fleshy stem and decurrent gills. *C. clavipes* (Fries) Kummer has a smoky brown, top-shaped cap, fragile stem which also has a distinct swelling at its base, and strong rather unpleasant smell. *C. nebularis* (Fries) Kummer is similar, but is pale cloudy grey, has a less fragile stem and a fairly pleasant smell. This species if often covered in a bloom which develops further as the fruit-body deteriorates. The agaric *Volvariella surrecta* (Knapp) Singer is a rare parasite of *C. nebularis* (see p. 247) and it has been suggested that this bloom may in fact belong to this species. How-ever, I have on several occasions tried to encourage the bloom to reproduce by keeping hoary looking fruit-bodies of *C. nebularis* in a damp-chamber, but as yet I have never been successful.

Nevertheless, it is an exercise which would be of great interest to continue and a source of great excitement if the small pink-spored agaric were produced. *C. fragrans* (Fries) Kummer is a small, sweetly aromatic-smelling species found in frondose woods, and *C. langei* Hora, is a mealy-smelling species of conifer plantations.

Illustrations: F 16a; Hvass 55; LH 95; WD 16².

Plate 19. Fleshy fungi: Spores white and borne on gills

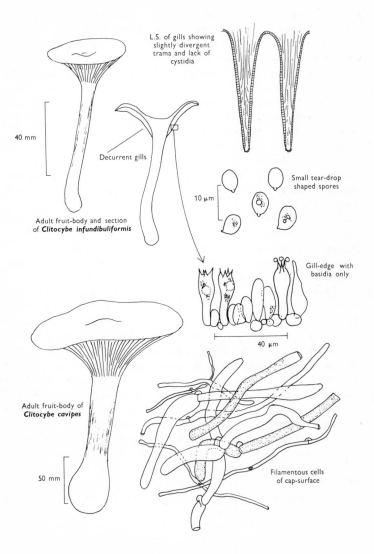

L.S. of gills showing slightly divergent trama and lack of cystidia

Decurrent gills

40 mm

Adult fruit-body and section of **Clitocybe infundibuliformis**

10 μm

Small tear-drop shaped spores

Gill-edge with basidia only

40 μm

Adult fruit-body of **Clitocybe cavipes**

50 mm

Filamentous cells of cap-surface

Hebeloma crustuliniforme (St Amans) Quélet   Fairy-cake mushroom
Cap: width 40–80 mm. *Stem:* width 8–12 mm; length 38–85 mm.
Description:

Cap: pale yellow buff or pale tan with a distinct reddish buff or
   cinnamon-brown tint, darkening only slightly with age; smooth,
   at first tacky to the fingers, but then dry and shiny at centre,
   convex and hardly expanding.

Stem: cylindrical or slightly swollen towards the base, whitish and with a
   flush of pinkish buff at apex, and covered all over in small, white scales.

Gills: sinuate, crowded, pale clay-colour or buff, but finally dull dark
   yellow ochre except for the distinct white margins which are beaded
   in wet weather with droplets of liquid.

Flesh: whitish with a very strong smell of radishes.

Spore-print: dark clay-colour.

Spores: long, slightly almond-shaped, pale brown under the microscope, distinctly warted and about 11×6 μm in size (10–12×6–7
   μm).

Marginal cystidia: cylindrical to skittle-shaped with slightly to distinctly swollen apex.

Facial cystidia: absent.

Habitat & *Distribution:* Common in autumn on the ground by path-
   sides and in woodland clearings.

General Information: Recognisable by the uniform cinnamon or pinkish
   buff cap, white woolly scales on the stem and distinctive, strong
   smell of radish. There is some evidence that this species may on
   occasions be mycorrhizal; further field studies are required.

   There are several closely related fungi which are difficult for the
amateur to differentiate from *H. crustuliniforme*; there is no doubt
that there are several species present in the British Isles which do
not appear in the Check List of British Agarics & Boleti; in fact,
it would appear that there are several yet to be described as new to
science. Although individual species are fairly difficult to delimit,
the genus *Hebeloma* itself is easily recognised, most members being
medium sized with brown sinuate gills, whitish, yellowish, or pinkish,
i.e. pale, caps and white-powdered stems. The word 'crustulin'
which appears in the Latin name of *H. crustuliniforme* is itself from
the Latin and means small cake, referring to the cap-shape, which
remains fairly constant throughout the fungus' growth. The common
name is derived from this also.

Plate 20. Fleshy fungi: Spores dull brown and borne on gills

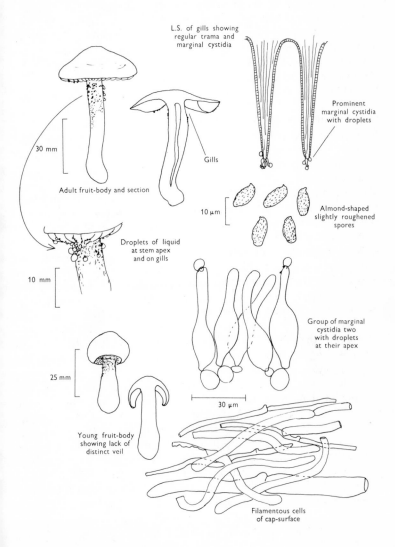

L.S. of gills showing regular trama and marginal cystidia

Prominent marginal cystidia with droplets

Gills

30 mm

Adult fruit-body and section

10 µm

Almond-shaped slightly roughened spores

Droplets of liquid at stem apex and on gills

10 mm

Group of marginal cystidia two with droplets at their apex

25 mm

30 µm

Young fruit-body showing lack of distinct veil

Filamentous cells of cap-surface

Inocybe geophylla (Fries) Kummer   Common white inocybe
Cap: width 10–25 mm. *Stem:* width 3–6 mm; length 30–50 mm.
Description:

Cap: conical with incurved margin then bell-shaped and retaining a
distinct umbo even when mature, silvery white then ivory and
finally pale tan particularly centrally and silky fibrillose throughout.

Stem: slender, cylindrical but for a small swelling at the base, silky
and shining with a few fibrils from a former cortina which may be
brownish due to spores adhering to it at maturity.

Gills: adnexed to free, crowded, pale ochraceous becoming clay-
coloured.

Flesh: white with smell of newly dug potatoes, strong when fresh.

Spore-print: clay-colour.

Spores: medium sized, ellipsoid or slightly French-bean-shaped,
smooth, yellow-brown under the microscope and 9–11 × 4–5 μm
in size.

Marginal and facial cystidia: flask- to spindle-shaped with distinctly
thickened walls and frequently ornamented with crystals apically.

Habitat & Distribution: Common in troops in woodland clearings, by
pathsides or on the edges of ditches bordering woods.

General Information: This fungus is easily recognised by the very pale
uniform colour, the colour of the spore-print, silky umbonate cap
and small size. The cortina connects the cap-margin and the stem
and consists of a cobwebby structure which collapses at maturity.

A violet coloured variety, var. *lilacina* Gillet is frequently found,
in fact, even accompanying var. *geophylla*; it differs only in the lilac-
colour of the cap and stem. *I. geophylla* is a member of the very large
genus *Inocybe*, further members of which will be dealt with later
(see p. 238).

The genus is well defined with dull-yellow spore-print, well
differentiated sterile cells on the gill-edge (and often on the gill-
face) and the cobweb–like veil, or cortina, stretching from the cap-
margin to the stem and easily observed in young specimens. The
genus is split into three distinct groups: those with smooth spores,
those with nodulose spores and those with subglobose spores ornamented with long projections. *I. geophylla* is included in the first
group. The group which includes the nodulose-spored members
has been elevated to the rank of genus by some authors, i.e. *Astrosporina*—a name referring to the spore-shape eg., *I. asterospora*.

Illustrations: F 13a (too blue); LH 155; NB 139[5]; WD 65[4].

Plate 21. Fleshy fungi: Spores dull brown and borne on gills

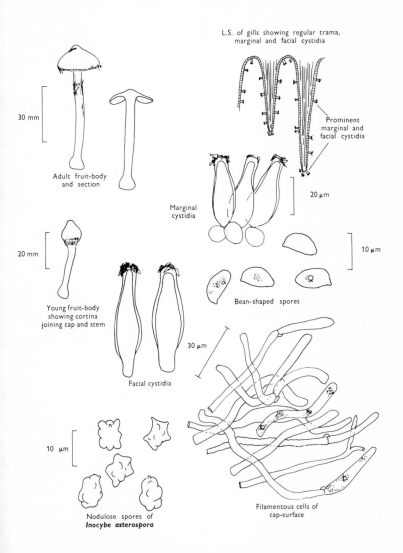

30 mm

Adult fruit-body
and section

20 mm

Young fruit-body
showing cortina
joining cap and stem

Facial cystidia

10 μm

Nodulose spores of
Inocybe asterospora

L.S. of gills showing regular trama,
marginal and facial cystidia

Prominent
marginal and
facial cystidia

20 μm

Marginal
cystidia

10 μm

Bean-shaped spores

30 μm

Filamentous cells of
cap-surface

Laccaria laccata (Fries) Cooke   Deceiver
Cap: width 12–28 mm. *Stem:* width 4–8 mm; length 15–60 mm.
Description:

Cap: hygrophanous, reddish brown or brick-colour becoming ochraceous on drying, but can be rapidly returned to the original colour by placing on the top a drop of water which is rapidly absorbed; fragile, convex at first then flattened or depressed about centre, smooth or surface scaley, striate at margin when moist.

Stem: similarly coloured to the cap, fibrous, cylindrical, tough and usually with white woolly base.

Gills: adnate with or without a decurrent tooth, thick, distant and pinkish or pale reddish-brown, powdered with white when mature.

Flesh: red-brown, soft in the cap and fibrous in the stem.

Spore-print: pure white.

Spores: medium sized, hyaline under the microscope and spherical, 7–8 μm in diameter and beautifully spiny.

Marginal and facial cystidia: absent.

Habitat & Distribution: Common in troops in woodland, copses, on heaths; in fact it may be found in nearly all possible habitats.

General Information: This is a very common agaric which in the future will probably be split into several distinct species; unfortunately it is as variable as it is common, hence the common name 'deceiver'; it is often mistaken at first glance for many other species quite unrelated. I have seen even the most experienced mycologist pick up rather unfamiliar specimens of *Laccaria laccata* in mistake for a species of *Lactarius* or a species of *Collybia*, etc. I would hate to say more because I have been 'deceived' myself on more than one occasion. *L. laccata* appears to be a composite species, but because of the difficulty in defining some of the characters the splitting of the species has not as yet been satisfactorily solved. The smell, however, may well give a clue for some specimens smell very strongly of radish whilst others are odourless.

L. proxima (Boudier) Patouillard, differs in having ellipsoid spores; it is larger in stature and is common in wet places.

L. amethystea (Mérat) Murrill, differs in the deep violet or amethyst-colour of the fruit-body and commonly grows in shaded woods.

L. bicolor (Maire) P. D. Orton, which is less frequent, has lilaceous gills and violaceous mycelium at the base of the stem.

Illustrations: Hvass 66; NB 133[1]; WD 20[2].

Plate 22. Fleshy fungi: Spores white and borne on gills

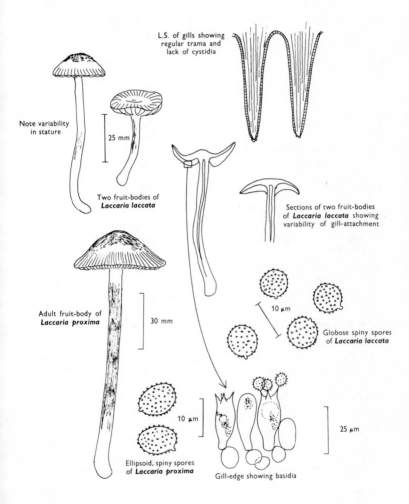

L.S. of gills showing regular trama and lack of cystidia

Note variability in stature

25 mm

Two fruit-bodies of *Laccaria laccata*

Sections of two fruit-bodies of *Laccaria laccata* showing variability of gill-attachment

Adult fruit-body of *Laccaria proxima*

30 mm

Globose spiny spores of *Laccaria laccata*

10 μm

10 μm

25 μm

Ellipsoid, spiny spores of *Laccaria proxima*

Gill-edge showing basidia

Mycena sanguinolenta (Fries) Kummer   Small bleeding mycena
Cap: width 10–17 mm. *Stem:* width 2–4 mm; length 50–80 mm.
Description:
Cap: bell-shaped or conical expanding only slightly with age and so remaining umbonate, reddish-brown, striate to the margin from the darker apex and blotched age with red-brown spots.
Stem: pale reddish brown, very slender, fragile, woolly at the base and exuding a red-brown juice when broken.
Gills: adnate, fairly distant, whitish to flesh-colour with a dark red-brown edge and not noticeably becoming blotched with red-brown.
Flesh: with no distinctive smell, reddish-brown and very thin.
Spore-print: white.
Spores: medium sized, hyaline, ellipsoid to pip-shaped, smooth about 10 μm long (9–10 × 4–5 μm) and becoming bluish grey when mounted in solutions containing iodine.
Marginal cystidia: awl-shaped, pointed at the apex, swollen below and filled with dark red-brown contents.
Facial cystidia: absent.
Habitat & *Distribution:* Solitary or in small groups on poorly kept lawns, in woods and copses; it is particularly frequent in the beds of needles found in pine woods.
General Information: This fungus is easily recognised by the slender habit, reddish juice exuded when broken and habitat preferences. *Mycena haematopus* (Fries) Kummer is larger and grows in tufts on wood, but also has a red-brown juice which, however, spots the gills. Another very common species of Mycena is *M. galopus* (Fries) Kummer which has a greyish or brownish cap and exudes a milk-like juice. The related *M. leucogala* (Cooke) Saccardo is almost black (see p. 216). These agarics exuding juice when broken have a flesh composed of filaments, a very different flesh-structure to species of *Lactarius* (see p. 50) and although their spores are amyloid they do not turn blue-black in iodine because of the presence of amyloid crests and warts. There are few additional species of agaric which exude a milk-like liquid, but the majority of these are tropical or subtropical. The second names or epithets for the four species mentioned above all refer to the 'latex'—sanguinolenta—bleeding, *haematopus* blood-foot; *galopus*, milk-foot and *leucogala*, white milk. For notes on *Mycena* one is referred to p. 68 describing *M. galericulata* (Fries) S. F. Gray.
Illustrations: WD 28[4].

Plate 23. Fleshy, milking fungi: Spores white and borne on gills

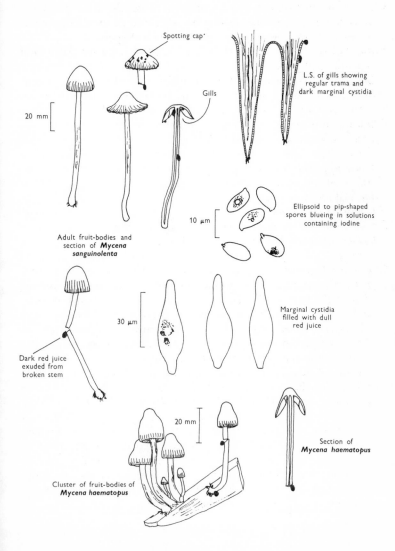

Spotting cap'

Gills

L.S. of gills showing
regular trama and
dark marginal cystidia

20 mm

Adult fruit-bodies and
section of **Mycena
sanguinolenta**

10 μm

Ellipsoid to pip-shaped
spores blueing in solutions
containing iodine

30 μm

Marginal cystidia
filled with dull
red juice

Dark red juice
exuded from
broken stem

20 mm

Section of
Mycena haematopus

Cluster of fruit-bodies of
Mycena haematopus

Collybia maculata (Fries) Kummer   Spotted tough-shank
Cap: width 80–130 mm. *Stem:* width 5–20 mm; length 50–158 mm.
Description:

Cap: white but soon becoming spotted with reddish-brown, finally
   cream-colour with red-brown blotches, convex then becoming
   flattened, fleshy, firm and tough.

Stem: white becoming streaked red-brown, thickest in the middle,
   longitudinally furrowed or striate and often narrowed downwards
   into a long irregular root embedded in the deep litter.

Gills: very crowded, cream-coloured, becoming spotted red-brown
   with age.

Flesh: with pleasant smell, white and fibrous in the stem.

Spore-print: pinkish cream-colour.

Spores: small, almost spherical, hyaline under the microscope, about
   5 μm in diameter (4–5 × 5 μm) and not blueing when placed in solutions containing iodine.

Marginal and facial cystidia: absent.

Habitat & *Distribution:* Common in troops in woods, particularly
   beech but also found in pine woods and on heaths.

General Information: Easily recognised by the crowded, narrow, cream
   coloured gills and the cap being entirely white when young, but which
   rapidly becomes spotted red-brown as it develops. 'Maculatus'
   means spotted and refers to the red-brown blotches which develop
   irregularly on the cap, stem and gills as the fruit-body matures.

   The genus *Collybia* is characterised by the fruit-body being tough,
   the cap-margin incurved at first and the spore-print white or whitish.
   The common fungus *C. maculata* has always been assumed to have
   a white spore-print but if a cap is placed on a piece of white paper
   gills-down and left for twelve hours there is a suprise in store for
   the careful observer.

Illustrations: F 15a; Hvass 77; LH 101; NB 103[4]; WD 21[2].

Plate 24. Fleshy fungi with tough stem: Spores white to cream and borne on gills

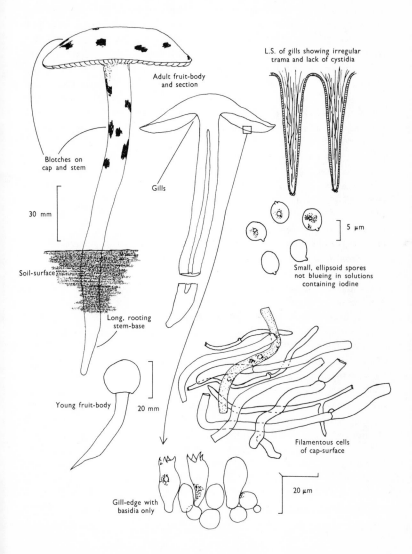

L.S. of gills showing irregular trama and lack of cystidia

Adult fruit-body and section

Blotches on cap and stem

Gills

30 mm

Soil-surface

5 µm

Small, ellipsoid spores not blueing in solutions containing iodine

Long, rooting stem-base

Young fruit-body

20 mm

Filamentous cells of cap-surface

Gill-edge with basidia only

20 µm

The specialised substrates of certain species of *Marasmius* and related genera

A whole series of very small fungi are found in woodland communities which appear to be closely related one to another because their caps are usually tough, although membranous, dry rapidly yet do not decay, and, moreover, revive on remoistening. Their gills are also rather tough and their spores always white in mass. They are placed in the genus *Marasmius*. *Collybia* or *Marasmius peronatus* (Fries) Fries the 'wood woolly foot' is one of our larger more familiar agarics related to this group, but whereas it grows on all kinds of leafy detritus, even wood, these small fungi appear to be very specific to the substrate on which they grow.

M. androsaceus (Fries) Fries grows both on heather and on pine-needles (see p. 231).

Cap: whitish or pinkish buff.

Stem: black and hair-like.

Spores: pip-shaped and 7–9 × 3–4 μm.

M. buxi Fries grows on box leaves.

M. epiphylloides (Rea) Saccardo & Trotter grows on ivy leaves.

M. graminum (Libert) Berkeley grows on grass stems.

Cap: red-brown.

Stem: dark brown.

Spores: pip-shaped, 8–12 × 4–6 μm.

M. hudsonii (Fries) Fries grows on holly leaves.

M. perforans (Fries) Fries grows on pine needles (now placed in the genus *Micromphale*).

M. undatus (Berkeley) Fries grows on bracken stems.

Cap: reddish brown or greyish and wrinkled.

Spores: egg-shaped, 8–9 × 6–7 μm.

Except for their rather special requirements as to substrate preference, these species have in common small size, rather tough horny stems and cap composed of erect ornamented cells.

Several agarics which grow on cones have also been placed in *Marasmius*. They are frequent in spring and early summer the fruit-bodies being attached by a very long rooting stem and cord of fluffy hyphae to buried cones in conifers. The biology of these fungi is still unknown, but the cones to which they are attached are always closed yet buried often several inches beneath the surface of the soil.

Plate 25. Fleshy fungi with wiry to tough stem: Spores white and borne on gills, fruit-body frequently reviving when moistened

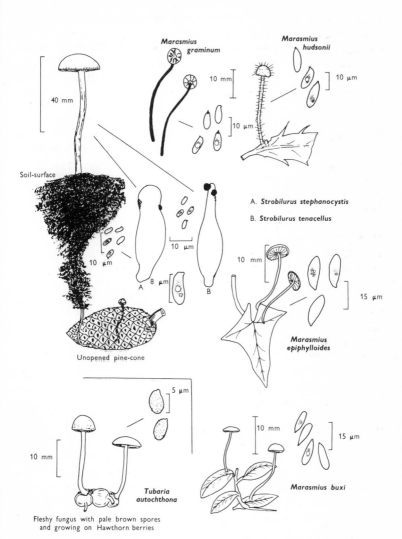

Marasmius graminum

Marasmius hudsonii

40 mm

10 mm

10 μm

Soil-surface

10 μm

A. *Strobilurus stephanocystis*

B. *Strobilurus tenacellus*

10 μm

10 mm

A 8 μm B

15 μm

Marasmius epiphylloides

Unopened pine-cone

5 μm

10 mm

15 μm

10 mm

Tubaria autochthona

Marasmius buxi

Fleshy fungus with pale brown spores
and growing on Hawthorn berries

It is yet to be found whether the spores of the agaric infect the cones after they drop or whether the cones fall because they have become infected. How do the cones become so deeply buried? Are squirrels or rodents involved? All the species which grow on cones have brown or tawny caps and yellowish brown stems.

Strobilurus stephanocystis (Hora) Singer has cystidia with rounded heads and grows on pine-cones.

S. tenacellus (Fries) Singer has pointed cystidia and grows on pine cones.

S. esculentus (Fries) Singer has lance-shaped cystidia and grows on spruce cones.

Baeospora myosura (Fries) Singer is tough and pale-coloured and is similar in general characters to species of *Strobilurus*, but has amyloid spores and fruits on pine-cones in the autumn.

When discussing the specialised plant-substrates, such as cones, one must mention the small brown-spored, pale buff coloured agaric *Tubaria dispersa* (Persoon) Singer, or *Tubaria autochthona* (Berkeley & Broome) Saccardo, which grows on the ground under hawthorns, often in troops in summer and autumn, attached to old hardened hawthorn berries.

(ii) Agarics of Pastures and Meadows

(a) Agarics of rough and hill pastures

Hygrocybe pratensis (Fries) Donk   Butter mushroom
Cap: width 20–80 mm. *Stem:* width 5–12 mm; length 30–70 mm.
Description:
Cap: convex then expanding to become plano-convex with a broad
  low umbo, tan, pale russet or even yellowish buff throughout or
  slightly darker at the centre.
Stem: gradually thickened upwards, similarly coloured to the cap or
  paler if the cap is dark russet.
Gills: pale buff, deeply decurrent and often connected up at their
  bases by veins.
Flesh: buff or pale tan, thick and soft in the cap, slightly fibrous in the
  stem.
Spore-print: white.
Spores: medium-sized, ellipsoid to egg-shaped, hyaline under the
  microscope, 7–8 × 5 μm in size and not becoming bluish grey in
  solutions containing iodine.
Marginal and facial cystidia: absent.
Habitat & Distribution: Common in pastures or on heaths from early
  summer to late autumn.
General Information: A fungus easily recognised by the uniform buff-
  colour of the stem, cap and gills. As one might expect from the
  common name it is edible; it is held in high regard by many mush-
  room-pickers.
    Although 'pratensis' specifically means fields, reflecting the habitat
  of the fungus, this and related species can also be found on heaths and
  pastures often intermixed and forming a most interesting flora. The
  following are perhaps the most commonly seen:
H. lacma (Fries) Orton & Watling and *H. cinerea* (Fries) Orton &
  Watling are similar in stature, but metallic grey in colour except for
  the persistently yellow stem-base in *H. lacma*.
H. subradiat (Secretan) Orton & Watling is flesh-coloured or brownish
  and *H. virginea* (Fries) Orton & Watling is white.
H. nivea (Fries) Orton & Watling and *H. russocoriacea* (Berkeley &
  Miller) Orton & Watling are much smaller, the former white and
  odourless and the latter off-white with a very strong smell of incense.
Illustrations: F 12[b]; Hvass 95; LH 77; NB 33[2]; WD 33[3].

Plate 26. Fleshy fungi: Spores white and borne on thick, waxy gills

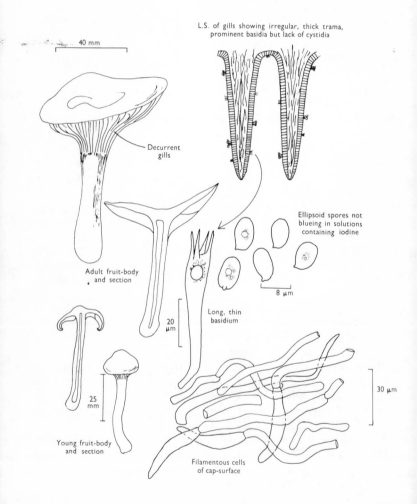

40 mm

L.S. of gills showing irregular, thick trama,
prominent basidia but lack of cystidia

Decurrent
gills

Ellipsoid spores not
blueing in solutions
containing iodine

8 μm

Adult fruit-body
and section

Long, thin
basidium

20
μm

30 μm

25
mm

Young fruit-body
and section

Filamentous cells
of cap-surface

Hygrocybe psittacina (Fries) Wunsche   Parrot hygrophorus
Cap: width 12–25 mm. *Stem:* width 3–8 mm; length 30–60 mm.
Description:
Cap: very slimy with colourless sticky fluid, deep bluish green when fresh, but becoming more and more ochraceous-orange with age or completely fading out to a yellow ochre, bell-shaped at first then expanded except for central umbo.
Stem: like the cap very slimy, apple-green or bluish green throughout but becoming ochraceous like the cap except at the apex which is persistently green.
Gills: adnate yellow or apricot-coloured, greenish towards their base, broad, distant and rather tough.
Flesh: whitish, tinged green in the cap and yellow or apricot-colour in the stem.
Spore-print: white.
Spores: medium-sized, hyaline, ellipsoid, not blue-grey in solutions containing iodine and 8–9 × 4–5 μm in size.
Marginal and facial cystidia: absent.
Habitat & *Distribution:* Common in grassland and hill-pastures, but it also occurs in copses and woodlands.
General Information: This fungus is easily recognised by the distinctive colours, but it is rather deceptive for the cap and the stem soon become faded; however, the green colouration persists at the apex of the stem and it is by this that in the faded state the fungus can still be identified. *H. laeta* (Fries) Kummer fades to similar colours but the cap is flesh-colour at first or sordid brown and the gills are flesh-coloured or greyish; it prefers upland pastures and heathland: its spores are smaller, being 5–7 × 4 μm.
Illustrations: F 12a; Hvass 92; LH 79; NB 33[6]; WD 34[5].

General notes on Hygrophori

Hygrophori are some of our most colourful groups of agarics, many are brightly coloured with caps in reds, greens, yellows, oranges, etc., the colour often accentuated by the usually slimy aspect. Traditionally the genus *Hygrophorus* has been split into three groups as follows:—

Limacium with slimy cap, adnate to decurrent gills and slimy or tacky stem which may also often be ornamented with dots, especially towards the top.

Camarophyllus with dry cap, smooth and fibrous stem and decurrent gills.

Hygrocybe with thin, fragile, sticky or moist cap, smooth fibrillose stem and gills varying from free to decurrent.

The last two sections have been joined together into the single genus *Hygrocybe* and all the members seem to be saprophytic or intimately associated with grassland communities. The first section *Limacium* now makes up the genus *Hygrophorus* and its members are thought to be mycorrhizal with trees, e.g. *H. hypothejus* (Fries) Fries with pine, the 'Herald of the winter' because it occurs at the end of the fungus season and *H. chrysaspis* Métrod, a whitish, sickly-smelling fungus under beech. Results from examining the anatomy of the gills appears to confirm these divisions. All the Hygrophori have a homogeneous flesh, white spores, central, fleshy stem and thick, waxy gills; microscopically this group of fungi can be recognised by the very long basidia.

The following are common examples of the genus Hygrocybe:—

H. calytraeformis (Berkeley & Broome) Fayod has a rose-pink, conical cap which expands to become upturned at the edge with age.

H. coccinea (Fries) Kummer has a bright scarlet cap which becomes yellow-ochre on drying and a yellow base to a scarlet stem.

H. conica (Fries) Kummer has an orange to red stem and sharply conical cap which turns blackish with age and whose gills when cut exude a clear watery liquid.

H. flavescens (Kauffman) Singer has a slimy, golden yellow cap and similarly coloured stem.

H. chlorophana is similar, but has a lemon-yellow cap and stem.

H. punicea (Fries) Kummer is a large and robust species, similar in colour to *H. coccinea* but with a white base to the stem.

H. unguinosa (Fries) Karsten has a smoky grey, very slimy cap and stem.

H. nitrata (Persoon) Wunsche is as dull coloured as *H. unguinosa*, but is not slimy, and in addition strongly smells of cleaning fluid or bleaching-powder. It is one of three dull coloured, strong bleaching-powder-smelling species found in Britain. *H. ovina* is another, but is darker than *H. nitrata* and becomes red when bruised or cut.

Plate 27. Fleshy brightly coloured fungi: Spores white and borne on thick, waxy gills

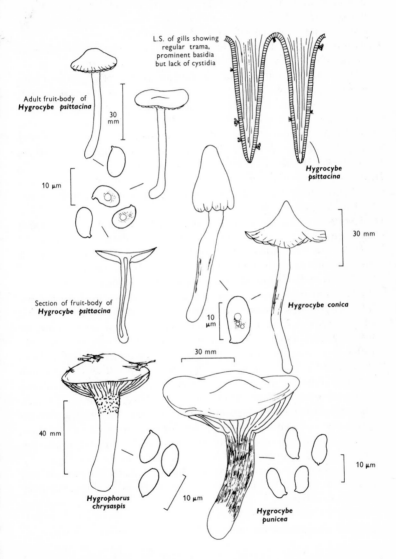

L.S. of gills showing regular trama, prominent basidia but lack of cystidia

Adult fruit-body of **Hygrocybe psittacina**

30 mm

10 µm

Hygrocybe psittacina

Section of fruit-body of **Hygrocybe psittacina**

Hygrocybe conica

30 mm

10 µm

30 mm

Hygrophorus chrysaspis

40 mm

10 µm

Hygrocybe punicea

10 µm

H. metapodia (Fries) Moser has a sooty brown fibrillose-streaky cap and stem. The gills are distant and grey, and the fruit-body may reach up to 100 mm across. It is probably the biggest of our native species of *Hygrocybe*.

For completion examples of *Hygrophorus* include:

H. bresadolae Quélet has a slimy orange-yellow cap, yellow gills and yellow, slimy, smooth stem. It is found under larch trees.

H. chrysaspis Métrod has ivory white cap, stem and gills which soon become flushed with rust-brown and finally the whole fruit-body becomes red-brown. The stem is slimy and white dotted at the apex. It grows in beech woods.

H. hedrychii Velenovsky has a slimy cream-coloured cap flushed with pale peach colour. The gills and stem are cream and the latter slimy and dotted at the top. It is found in pine woods.

H. hypothejus (Fries) Fries has an olive-brown slimy cap, yellow stem and gills; the stem is slimy and smooth. It is found in pine woods and under pines on heaths.

H. pustulatus (Persoon) Fries has an ash-grey cap brownish towards its centre, viscid white stem with dark grey dots at the apex and white gills. *H. agathosmus* (Secretan) Fries is similar, but smells strongly of bitter almonds. Both species are found in plantations.

Species of the genus *Hygrophorus* are infrequently encountered in Britain, although twenty species are recorded for the British Isles. They are ecologically distinct from members of the genus *Hygrocybe* in preferring woodland communities to grassland areas; they are probably mycorrhizal. The anatomy of the fruit-body is also rather different to that found in *Hygrocybe*; the gill-trama is bilateral as in *Leccinum* (p. 27), *Suillus* (p. 28), *Boletus* (p. 31), *Chroogomphus* (p. 36), *Paxillus* (p. 38) and *Amanita* (p. 54). Members of the genus *Hygrocybe* have regular to irregular gill-tramas. In fact, although both genera are united into a single family, the Hygrophoraceae is based on one character common to both, i.e. the long basidium; there is every indication that the genus *Hygrocybe* has greater affinity to *Omphalina* in the Tricholomataceae (p. 232).

Surprisingly enough in North America many of our familiar grassland species including *H. pratensis* are to be found in deep shaded woodland!

Angular, pink-spored agarics—Rhodophyllaceae

The name of the family refers to the pink gills and it unites all those fungi with a salmon-pinkish buff spore-print and whose spores are angular in all optical sections. There are a few agarics, e.g. *Clitopilus prunulus* (Fries) Kummer with ridged spores which appear angular in end-on view, but which are ellipsoid in both side and face views and so are considered less related.

The family *Rhodophyllaceae* by some authorities contains one genus *Rhodophyllus*, more correctly called *Entoloma*; in the British Isles five constituent genera are recognised, but they will have to be more critically defined to make a more meaningful classification. At the moment, many of the species are poorly documented and it would appear that anatomical studies will assist in the future in the recognition of species-groups.

If one selects the eight most distinctive shaped spore-types exhibited in members of this family, then when their spores are examined side-on a feature is available for correlation with the traditional field characters, such as cap scaliness and gill-attachment. The most distinctive spore-shape is Type G, found in *Nolanea staurospora* Bresadola, which is probably the most common and widespread species of the family. It grows in woodlands, grassland and on lawns and will be dealt with later (p. 122). The other spore types are illustrated and range from irregularly rhomboid to elongate angular.

The majority of the members of this group grow in grassland, hill-pastures and meadows and distinct communities containing members of this family and of the *Hygrophoraceae* can be recognised. It is not proposed to deal in detail with any individual members because they can be so easily confused with each other by the specialist let alone by the amateur.

However, the genera as at present accepted are as follows:—

1. ***Entoloma*** in its original sense contains agarics with fleshy caps, fibrous stems and sinuate or adnexed gills, e.g. *Ent. clypeatum* (Fries) Kummer with grey to yellow-brown cap, found growing with members of the apple and rose-family in the summer and early autumn. This genus corresponds to *Calocybe* in the white-spored agarics (p. 110).

2. **Leptonia** contains those agarics with rather thin caps whose margin is incurved, cartilaginous stems and adnate to adnexed, rarely decurrent, gills and whose cap flesh is indistinct from that of the stem, e.g. *Lept. serrulata* (Fries) Kummer with dark blue to violet-blue cap and dark blue edge to the gills. This genus approaches the tough-shanks (*Collybia*) in the white-spored genera (p. 90).

3. **Nolanea** is characterised by agarics with delicate caps, whose flesh is distinct from that of the stem and whose edge is straight and pressed against the fragile stem when young, and the adnexed or adnate, rarely decurrent, gills, e.g. *N. staurospora* (see p. 122). *N. cetrata* (Fries) Kummer with yellow-brown to tan-coloured cap is found from spring to autumn in conifer woodland, especially plantations. The genus corresponds to *Mycena* in the white-spored agaric genera (p. 68).

4. **Eccilia** is a small genus containing agarics with thin, membranous caps and distinctly decurrent gills, e.g. *E. sericeonitida* P. D. Orton with convex, then umbilicate, silky greyish brown cap. This genus corresponds to *Omphalina* in the white-spored agarics (p. 232).

5. **Claudopus** has three British representatives, all of which have a very small stem which may even be absent, e.g. *C. depluens* (Fries) Gillet grows on soil and *C. parasiticus* (Quélet) Ricken grows on old decaying fruit-bodies of woody fungi. This genus corresponds to *Pleurotellus* in the white-spored genera and to *Crepidotus* in the brown-spored genera (p. 77).

Plate 28. Fleshy fungi: Spores pinkish and angular and borne on gills - Rhodophyllaceae

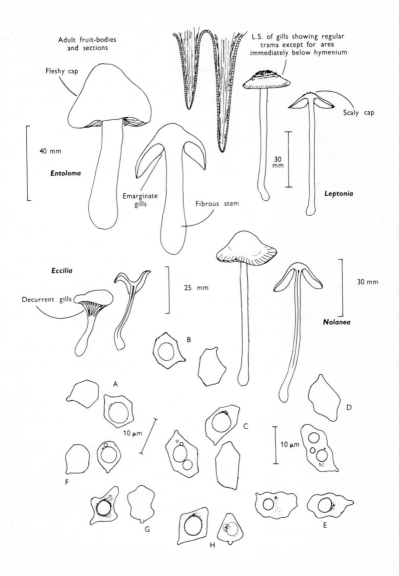

Adult fruit-bodies and sections

Fleshy cap

L.S. of gills showing regular trama except for area immediately below hymenium

Scaly cap

40 mm

Entoloma

30 mm

Leptonia

Emarginate gills

Fibrous stem

Eccilia

Decurrent gills

25 mm

30 mm

Nolanea

B

A

10 μm

10 μm

C

D

F

E

G

H

Cystoderma amianthinum (Fries) Fayod
Cap: width 15–35 mm. *Stem:* width 4–8 mm; length 15–30 mm.
Description:
Cap: pale ochraceous yellow to sand-colour, convex then expanded, with central umbo and often radially wrinkled-reticulate, covered completely in powdery granules when fresh but these gradually disappear with age or on excessive handling.
Stem: slender, white above a narrow, easily lost ring which is composed of floccose, ochraceous yellow granules which also clothe the lower part of the stem.
Gills: adnate, cream-coloured and crowded.
Flesh: yellowish with a strong smell of new-mown hay.
Spore-print: white.
Spores: small to medium sized, hyaline under the microscope, smooth, ellipsoid, 5–7 × 3–4 μm and becoming blue-grey when mounted in solutions containing iodine.
Marginal and facial cystidia: absent.
Habitat & Distribution: Frequently found amongst grass on heaths, in hill-pastures and in woodlands from summer to autumn.
General Information: This fungus is recognised by the gill-attachment and the powdery-scurfy cap formed by the breaking up of an enveloping veil composed of thick-walled, rounded cells, similar to those on the surface of the stem.

This fungus was formerly placed in the genus *Lepiota* because of the ring but the veil in *Cystoderma amianthinum* is formed in quite a different way to the ring in the true parasol mushrooms. The gills are also adnate and not free as in the true species of *Lepiota* (see p. 112). *C. carcharias* (Secretan) Fayod is found under similar conditions, but is white or flesh-coloured. *C. cinnabarinum* (Secretan) Fayod is also found in short grass and moss, but has a cinnabar-red, floccose cap and *C. granulosum* (Fries) Fayod is yellowish brown with non-amyloid spores and adnexed gills.

Many authorities prefer to connect this small group of closely related species more to members of the *Tricholomataceae* (i.e. the family which contains the Wood Blewits (p. 131), *Mycena* (p. 68, etc.) than to the parasol mushrooms—*Lepiota* (p. 112).
Illustrations: Hvass 23; LH 129; NB 103[7]; WD 8[4].

Plate 29. Fleshy fungi: Spores white and borne on gills

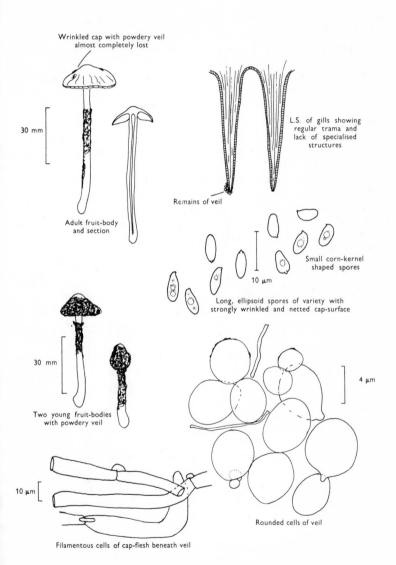

Wrinkled cap with powdery veil almost completely lost

30 mm

Adult fruit-body and section

Remains of veil

L.S. of gills showing regular trama and lack of specialised structures

Small corn-kernel shaped spores

10 μm

Long, ellipsoid spores of variety with strongly wrinkled and netted cap-surface

30 mm

Two young fruit-bodies with powdery veil

4 μm

10 μm

Rounded cells of veil

Filamentous cells of cap-flesh beneath veil

Hygrophoropsis aurantiaca (Fries) Maire   False chanterelle
Cap: width 25–70 mm. *Stem:* width 4–7 mm; length 25–50 mm.
Description:
Cap: bright orange-yellow or apricot, fleshy, soft, depressed at centre
and with wavy, incurved, slightly downy margin.
Stem: yellow at apex, rich red-brown or orange about the middle and
sometimes dark brown at the very base.
Gills: decurrent, deep orange, thin, crowded, repeatedly forked and
easily separable from the cap-tissue.
Flesh: yellowish, pale in the cap, darker in the stem.
Spore-print: white.
Spores: medium sized, hyaline under the microscope, ellipsoid or
pip-shaped, smooth, 7–8 × 4 μm and red-brown when mounted in
solutions of iodine.
Marginal and facial cystidia: absent.
Habitat & *Distribution:* Common in woodlands, particularly with
pines, and on heaths or in rough hill-pastures.
General Information: This fungus is recognisable by the orange or
yellow cap and stem and the decurrent gills. It was formerly placed
in *Cantharellus* because of the colours, white spores and the decurrent
gills, but it really differs in many other respects. It is true, however,
that it is frequently confused with the true Chanterelle (*Cantharellus
cibarius* Fries, p. 162) by those who do not inspect their specimens
carefully. The gills are thin, plate-like as in other agarics and not
fold-like as in *Cantharellus* (see p. 162). The Chanterelle is edible
and sought after as a delicacy, but there are varying reports as to the
edibility of *Hygrophoropsis*. Certainly it is not of the best quality
and there is evidence for it causing upsets: therefore it is best to take
the name 'False Chanterelle' at face value and treat this fungus as
truely false; 'aurantiaca' means orange-coloured and refers to the
colour of the fungus.

A pale form is frequently collected, particularly in hill-pastures,
and is probably worthy of specific recognition. The cap is ochraceous
yellow to cream and the stem distinctly dark in the lower half.

There is some confusion as to the true position in classification of
this fungus. The anatomical details of the fruit-body parallel those
of *Paxillus involutus* (Fries) Fries (see p. 38) although the spore-
print is white. There is little doubt that future research will answer
this problem.
Illustrations: Hvass 183; LH 185; NB 103[1]; WD 16[3].

Plate 30. Fleshy fungi: Spores white and borne on gills

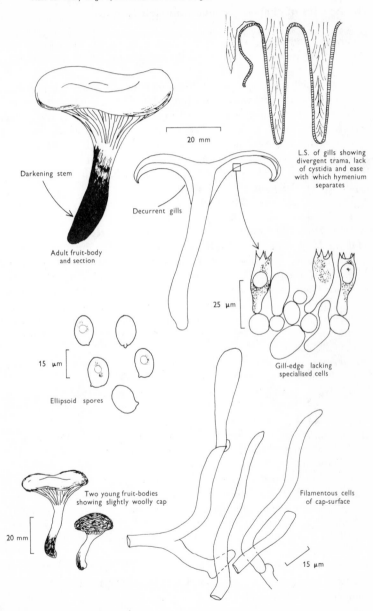

20 mm

L.S. of gills showing
divergent trama, lack
of cystidia and ease
with which hymenium
separates

Darkening stem

Decurrent gills

Adult fruit-body
and section

25 μm

15 μm

Ellipsoid spores

Gill-edge lacking
specialised cells

Two young fruit-bodies
showing slightly woolly cap

Filamentous cells
of cap-surface

20 mm

15 μm

(b) Agarics of chalk-grassland and rich uplands

Agaricus campestris Fries   Field mushroom
Cap: width 40–100 mm. *Stem:* width 12–20 mm; length 40–80 mm.
Description:
Cap: rounded then expanding to become plano-convex, fleshy with the
   margin incurved at first, initially pure white, but soon becoming cream-
   colour and at maturity streaked brownish particularly at the centre.
Stem: white with a simple, very thin, white ring which becomes
   brownish on rubbing and is easily lost with age or by handling.
Gills: free, pink but finally umber-brown at maturity.
Flesh: white, flushed reddish when cut especially in the stem.
Spore-print: cigar-brown, with hint of purple.
Spores: medium sized, ellipsoid or egg-shaped, smooth, small, 7–8 ×
   4–5 μm and dark brown under the microscope.
Marginal and facial cystidia: absent. Basidia 4-spored.
Habitat & Distribution: The field-mushroom grows amongst grass in
   pastures, etc., and also on old lawns where it may form fairy-rings.
General Information: This is the common wild, edible mushroom for
   which many people have in the past unwisely substituted many quite
   unrelated species. Deaths have often been caused by lack of careful
   observation when selecting wild fungi for the table; this only
   emphasises why 'white mushrooms' found in fields should not be
   casually eaten.
A. arvensis Secretan the 'Horse-mushroom' is also edible, but is
   much bigger (up to 180 mm), creamy white and bruises slightly
   yellowish on handling; it also has larger spores (7–10 × 5 μm),
   club-shaped cells on the gill-edge, gills commencing white and not
   pink, and the presence of a complex ring.
A. xanthodermus Genevier the 'Yellow-staining mushroom' has even
   smaller spores than the field mushroom, i.e. 5–6 × 4 μm and a
   rather strong, unpleasant smell; if eaten many people subsequently
   suffer from stomach-pains and this shows that even amongst those
   fungi which the scientist would call true mushrooms, i.e. those
   fungi in the genus *Agaricus*, there are some poisonous members.
   Thus it is always necessary to have wide experience before one
   collects fungi for eating and until this is achieved all specimens
   should be discarded.
Illustrations: Field mushroom—Hvass 163; LH 133; NB 31[6]; WD 71[2].
   Horse mushroom—Hvass 160; LH 135; WD 72[1]. Yellow-staining
   mushroom—Hvass 159; WD 71[3].

Plate 31. Fleshy fungi: Spores purple-brown and borne on gills

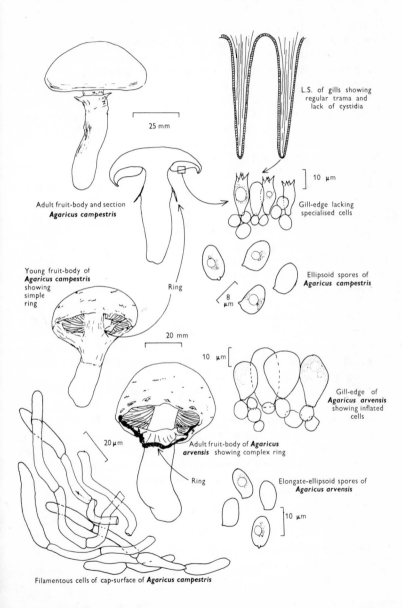

25 mm

L.S. of gills showing regular trama and lack of cystidia

10 μm

Gill-edge lacking specialised cells

Adult fruit-body and section **Agaricus campestris**

Ellipsoid spores of **Agaricus campestris**

8 μm

Young fruit-body of **Agaricus campestris** showing simple ring

Ring

20 mm

10 μm

Gill-edge of **Agaricus arvensis** showing inflated cells

20 μm

Adult fruit-body of **Agaricus arvensis** showing complex ring

Ring

Elongate-ellipsoid spores of **Agaricus arvensis**

10 μm

Filamentous cells of cap-surface of **Agaricus campestris**

Calocybe gambosum (Fries) Singer.   St George's mushroom
Cap: width 70–100 mm. *Stem:* width 15–25 mm; length 50–70 mm.
Description:

Cap: creamy white, ivory or light buff, slightly darker at the centre
  with age, fleshy, rounded and with wavy margin, finally expanding
  to become plano-convex; the margin is incurved and slightly downy
  at first.

Stem: firm, rather thick, white at the top, creamy or buff below and
  slightly downy when fresh.

Gills: sinuate to adnexed with a slight decurrent tooth, white to
  pale buff.

Flesh: with a very strong smell of meal, white and thick.

Spore-print: white.

Spores: small, ellipsoid, smooth, hyaline under the microscope, 5–6 ×
  3–4 μm and not becoming blue-grey with solutions containing
  iodine.

Marginal and facial cystidia: absent.

Habitat & *Distribution:* Found amongst grass in base rich pastures,
  often in fairly large rings from April to June and on golf-courses
  particularly those near the sea.

General Information: The common name refers to the early appearance
  of this agaric; St George's Day is April 23rd, and this mushroom
  is found about this time in favourable years, its fruiting often extending into early June, particularly if the fruiting is retarded by a
  cold and wet spring. It is easily recognised by the pale colour of the
  cap, strong mealy smell, but particularly by its appearance in spring.
  In each new year it is probably the first of the larger agarics to
  appear. This species will be found in most books under the genus
  Tricholoma, but differs from typical members of this group in the
  anatomy and chemistry of the gill-tissues.

  The Latin name 'gambosum' is derived from 'gamba' meaning a
  hoof and this reflects the shape of the fleshy cap as it pushes up
  through the grass. Another much older name is *Tricholoma georgii*
  (Fries) Quélet which was used by Clusius and is derived from the
  legend of St George.

Illustrations: Hvass 28; LH 83; WD 9².

Plate 32. Fleshy fungi: Spores white and borne on gills

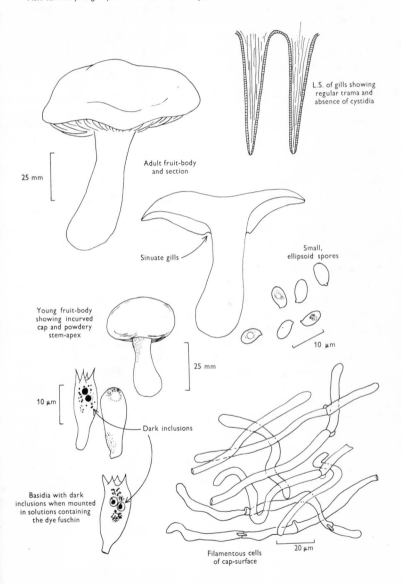

L.S. of gills showing regular trama and absence of cystidia

25 mm

Adult fruit-body and section

Sinuate gills

Small, ellipsoid spores

10 μm

Young fruit-body showing incurved cap and powdery stem-apex

25 mm

10 μm

Dark inclusions

Basidia with dark inclusions when mounted in solutions containing the dye fuschin

Filamentous cells of cap-surface

20 μm

Lepiota procera (Fries) S. F. Gray   Parasol mushroom
Cap: width 70–200 mm. *Stem:* width 12–20 mm; length 100–250 mm.
Description:

Cap: dull brown or greyish brown, oval or rounded at first, but later
   becoming bell-shaped, finally expanding but for the central umbo
   and the surface breaking up into shaggy scales.

Stem: straight, tapering upwards from a slightly bulbous base, felty
   at first but then the surface breaking up into small patches which
   finally resemble the pattern of a snake-skin; there is also a large,
   thick, white ring which is brown below and becomes loose on the
   stem.

Gills: remote, white, crowded and fairly broad.

Flesh: white, thin, soft.

Spore-print: white.

Spores: very long, ellipsoid with a germ-pore, hyaline under the
   microscope about 16 × 10 μm (14–17 × 9–12 μm), and becoming
   reddish brown in solutions containing iodine.

Marginal cystidia: variable, elongate balloon-shaped and hyaline.

Facial cystidia: absent.

Habitat & *Distribution:* Found from summer until mid-autumn, on
   the outskirts of copses, in fields, at edges of woodland or in woodland clearings; it is sometimes found in very large rings.

General Information: When this fungus first appears through the soil
   it resembles a drum-stick with the margin of the unexpanded cap
   tightly hugging the stem. It is an easily recognised fungus because
   of its straight and graceful stature with large cap and tall stem. It is
   one of our best edible fungi and cannot be confused with any other
   agaric. *L. rhacodes* (Vittadini) Quélet is not as elegant and has much
   smaller spores.

Illustrations: F 26a; Hvass 15; LH 125; NB 31[1]; WD 5[1].

Plate 33. Fleshy fungi: Spores white and borne on gills

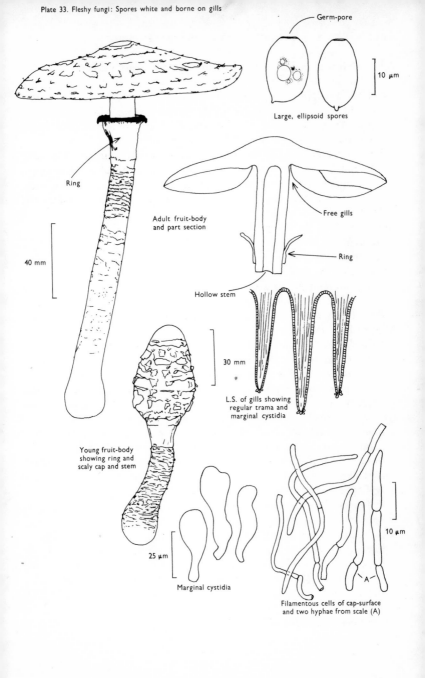

Germ-pore

Large, ellipsoid spores

10 µm

Ring

Adult fruit-body
and part section

Free gills

Ring

Hollow stem

40 mm

L.S. of gills showing
regular trama and
marginal cystidia

30 mm

Young fruit-body
showing ring and
scaly cap and stem

Marginal cystidia

25 µm

Filamentous cells of cap-surface
and two hyphae from scale (A)

10 µm

(c) Agarics of meadows and valley-bottom grasslands

Psilocybe semilanceata (Secretan) Kummer   Liberty caps
Cap: width 8–14 mm; height up to 18 mm. *Stem:* width 4–6 mm; length 50–70 mm.
Description:

Cap: sharply conical, in fact often with a very distinct apical point, never or very rarely becoming expanded, often fluted and puckered at the incurved margin, smooth, viscid, pale buff or clay colour, but soon flushed with greyish green at maturity and becoming free of the fibrils of veil which ornament the margin when young.

Stem: slender, tough and smooth, similarly coloured to the cap and sometimes blueing at the base when picked.

Gills: adnate to adnexed, crowded, purplish black except for white edge.

Flesh: white or pallid.

Spore-print: purple-brown.

Spores: long, ellipsoid, slightly lemon-shaped, smooth and with a distinct germ-pore at one end and 12–14 $\times$ 7 µm in size.

Marginal cystidia: bottle-shaped with an elongate tapering neck, with thin walls which at most become pale honey in solutions containing ammonia, unlike the cystidia of *Hypholoma* (p. 64).

Habitat & Distribution: Commonly growing amongst grass in fields near farm-yards, on heaths and by roadsides; often it occurs in small troops.

General Information: Psilocybe semilanceata is recognised by the uniquely shaped cap; 'semilanceata' means half spear-shaped, from the papilla at the top of the cap, giving it a pointed aspect. However, the common name is more descriptive and comes from the fact that these caps resemble the helmets worn by French soldiers in the early part of the century.

This fungus was once very isolated amongst British agarics, but now it has been united with a group of small purplish brown-spored fungi formerly placed in the genus *Deconica*. What is of more interest is the fact that unlike many British agarics the cap often does not expand fully in order to release the spores. In this way it allows mycologists to hypothesise on how certain of the enclosed, stalked Gastromycetes evolved in some of the desert regions of the world.

Illustrations: LH 149; NB 33[11]; WD 78[7].

Plate 34. Fleshy fungi: Spores purple-brown and borne on gills

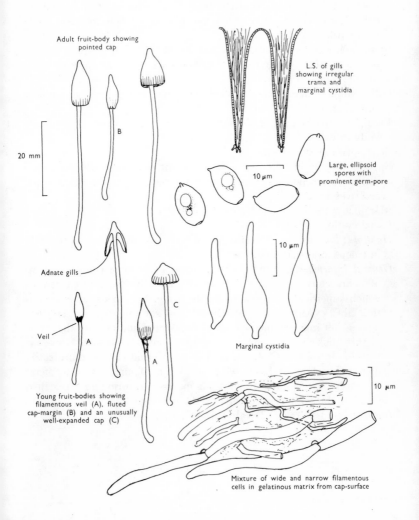

Adult fruit-body showing pointed cap

L.S. of gills showing irregular trama and marginal cystidia

20 mm

B

10 µm

Large, ellipsoid spores with prominent germ-pore

10 µm

Adnate gills

C

Veil

A

A

Marginal cystidia

Young fruit-bodies showing filamentous veil (A), fluted cap-margin (B) and an unusually well-expanded cap (C)

10 µm

Mixture of wide and narrow filamentous cells in gelatinous matrix from cap-surface

Conocybe tenera (Fries) Fayod   Brown cone-cap
Cap: width 10–20 mm. *Stem:* width 3–6 mm; length 70–100 mm.
Description:
Cap: very hygrophanous, sand colour, orange-yellow or ochraceous
   brown tinted cinnamon when fresh but drying uniformly yellow-
   ochre, thin, fragile, striate when moist, but soon non-striate as water
   is lost from the cap.
Stem: tall, slender and similarly coloured to the cap, straight, fragile,
   minutely striate from the top to bottom with what appears to be
   minute powdery granules.
Gills: adnate then becoming free, crowded, ochraceous and finally
   cinnamon-rust in colour.
Flesh: russet when moist but rapidly becoming yellowish as the fruit-
   body dries.
Spore-print: rust-brown.
Spores: long, ellipsoid, with thick, bright yellow-brown walls and
   distinct germ-pores at their ends when seen under the microscope,
   and over 10 μm in length (11–12 × 6 μm.)
Marginal cystidia: pinheaded or skittle-shaped.
Facial cystidia: absent.
Habitat & *Distribution:* This fungus grows in ones and twos, more rarely
   in troops amongst grass.
General Information: This is one member of a whole complex group
   of ochraceous, brown, tawny or cinnamon-brown capped agarics
   which superficially appear to be the same, but on closer examination the expert can split them into several distinct species.
   The use of microscopic characters is essential and outside the scope
   of this book or the ordinary mushroom-picker's manual. However,
   the closely related *C. lactea* (J. Lange) Métrod can be more easily
   distinguished for it has a white or cream-coloured cap and stem.
   It also has larger broadly ellipsoid spores, measuring 12–14 × 6–9
   μm, but the same shaped cells on the gill-edge.
Illustrations: LH 153; NB 35[4]; WD 68[2].

Plate 35. Fleshy fungi: Spores brown and borne on gills

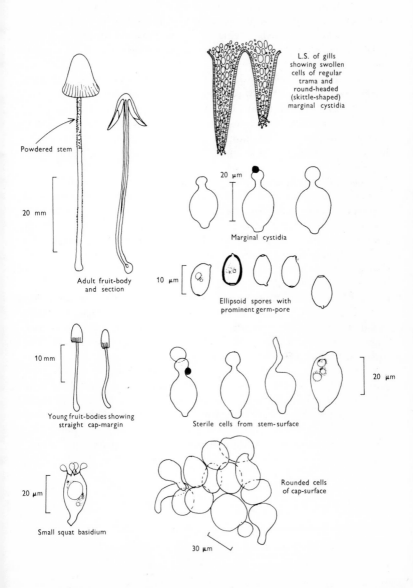

Powdered stem

20 mm

Adult fruit-body and section

L.S. of gills showing swollen cells of regular trama and round-headed (skittle-shaped) marginal cystidia

20 μm

Marginal cystidia

10 μm

Ellipsoid spores with prominent germ-pore

10 mm

Young fruit-bodies showing straight cap-margin

20 μm

Sterile cells from stem-surface

20 μm

Small squat basidium

Rounded cells of cap-surface

30 μm

(d) Fairy-ring formers

Many agarics grow in circles, but not all of them produce zones in the vegetation. It is the distinct zonation caused by the 'fairy-ring champignon' *Marasmius oreades* (Fries) Fries and related fungi which have given rise to the name of Fairy-ring and which resulted in the foundation of many folk tales.

A fairy-ring can be divided into four distinct zones, a central zone of fairly normally developed vegetation on the outside of which is a green, actively growing zone of grass; outside this is a zone composed of brown or dead vegetation. The outermost zone again appears to be far more lush than the normal grass in the vicinity and it is in this last zone that the fruit-bodies of the fungus causing the pattern appear.

A generalised explanation of the zoning appears to be as follows:—

In the outermost zone the actively growing mycelium decomposes soil constituents and liberates nitrogenous material which is in turn taken up by the plant roots nearby and utilised for their growth. In the penultimate zone the grass is dead, probably not caused by a direct parasitic attack but by the mycelial threads filling the air-spaces in the soil and so inhibiting water flow. This destruction of the delicate balance of water and air found in any soil induces drying out and gradual death of the plants whose roots permeate the soil. Behind the dead-zone is vegetation which shows increased vigour apparently due to plant-nutrients being released by the decaying mycelium and plant-material, whose death has been caused by the presence of the fungus. The innermost zone is not so stimulated.

With nothing more than graph and tracing paper, a tape-measure, note-book and pencil, pieces of cane about four inches long, and coloured dye or indian ink, it is exciting to assess the annual radial growth of fairy-rings and to correlate these with environmental conditions. This can be carried out on a school lawn or on a home lawn; the method and further experiments are given in Appendix iii.

Plate 36. Fairy-ring fungus—*Marasmius oreades*

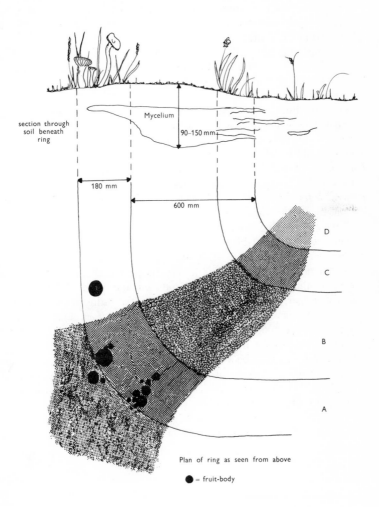

section through soil beneath ring

Mycelium

90–150 mm

180 mm

600 mm

D

C

B

A

Plan of ring as seen from above

● = fruit-body

Marasmius oreades (Fries) Fries   Fairy-ring champignon
Cap: width 25–60 mm. *Stem:* width 5–9 mm; length 30–80 mm.
Description:

Cap: pinkish tan with slight flush of brown at centre, hygrophanous
  and drying out buff-coloured or clay-coloured, convex at first then
  expanding to become plane, but for an obtuse umbo which is
  retained at the centre.

Stem: pale buff, tough, flexible and smooth.

Gills: adnexed, pale cream colour or pinkish buff and fairly distant.

Flesh: whitish or pinkish tan, smelling of cherry laurel (cyanic).

Spore-print: white.

Spores: medium sized, hyaline, pip-shaped, smooth, not staining
  bluish grey when mounted in solutions containing iodine and about
  10×6 μm in size (9–11×5–6 μm).

Marginal and facial cystidia: absent.

Habitat & Distribution: This agaric is very common from May to
  October on lawns and grass-verges.

General Information: M. oreades forms well developed fairy-rings, and
  is easily recognised by its tough nature, pale colours and ability to
  revive after having been dried. This ability to revive in moist weather
  even after the fruit-body has been dried by the sun or wind is a
  character which was used to distinguish members of the genus
  Marasmius. However, this is a very subjective character and since
  microscopic techniques were introduced and used widely in the
  study of agarics the genus has been delimited rather more critically.
  Marasmius is close to *Collybia* (p. 90), in fact many species
  appear in one book in one genus and in another book in the second
  genus; *M. oreades* itself is not a typical member of the genus.
  Marasmius seems to be a much more important genus in the tropical
  and subtropical regions of the world; we have already mentioned
  how some of the small species of *Marasmius* in Europe grow only
  on leaves of a particular plant (see p. 92). *M. androsaceus* (Fries)
  Fries (see p. 231) is the horse-hair fungus.

Illustrations: F 19a; Hvass 81; LH 115; NB 35[1]; WD 24[10] (not very
  good).

Plate 37. Fleshy fungi reviving when moistened even after drying; Spores white and borne on gills

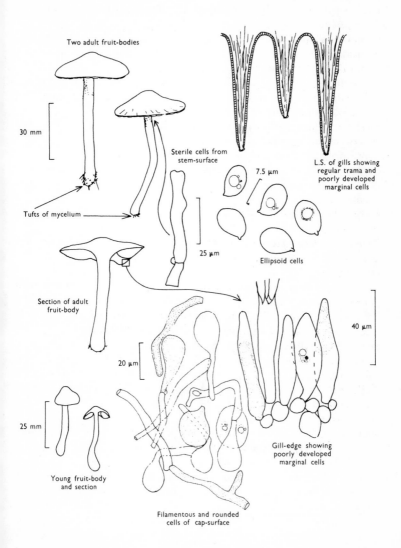

Two adult fruit-bodies

30 mm

Tufts of mycelium

Sterile cells from stem-surface

7.5 μm

25 μm

Ellipsoid cells

L.S. of gills showing regular trama and poorly developed marginal cells

Section of adult fruit-body

40 μm

20 μm

25 mm

Young fruit-body and section

Filamentous and rounded cells of cap-surface

Gill-edge showing poorly developed marginal cells

(e) Agarics of urban areas—lawn and parkland agarics

Nolanea staurospora Bresadola
Cap: width 20–40 mm. *Stem:* width 3–5 mm; length 45–70 mm.
Description:
Cap: bell-shaped at first then expanded, hygrophanous, date-brown, striate when moist but pale fawn or tan and non-striate when dry, and usually becoming quite silky-shiny.
Stem: slender, fragile, greyish brown, silky fibrillose-striate and shiny.
Gills: almost free, crowded and pale greyish brown when young, but finally flesh coloured.
Flesh: brownish and smelling very strongly of meal when cut or broken between the fingers.
Spore-print: salmon-pink with flush of cinnamon.
Spores: medium sized, fawn under the microscope, star-shaped with 4–6 prominent angles, 9–10 × 7–9 μm, smooth and with no germ-pore.
Marginal and facial cystidia: absent.

Nolanea sericea (Mérat) P. D. Orton   Silky nolanea
Cap: width 25–40 mm. *Stem:* width 5–9 mm; length 25–50 mm.
Description:
Cap: convex then flattened or with slight umbo, umber-brown with a greyish cast which becomes accentuated as the cap dries out and finally becoming silky-shiny; the margin is incurved and striate at first but on expanding it becomes non-striate with time.
Stem: short, fibrillose, greyish brown, shining and white at the base, very fragile and often snaps just above the soil-level when collected.
Gills: crowded, adnate and pale greyish brown then pinkish brown.
Flesh: with a strong smell of new meal, brownish becoming paler as it dries out.
Spore-print: salmon-pink.
Spores: medium sized, smooth, pale fawn under the microscope, angular almost cubic and 10–13 × 8–9 μm in size.
Marginal and facial cystidia: absent.
General Information: Nolanea staurospora is very common amongst grass, in many habitats such as on heaths, and in woodlands and copses, but it is particularly common in pastures and on lawns. It is difficult to separate from close relatives on field-characters,

Plate 38. Fleshy fungi: Spores pinkish and angular, and borne on gills

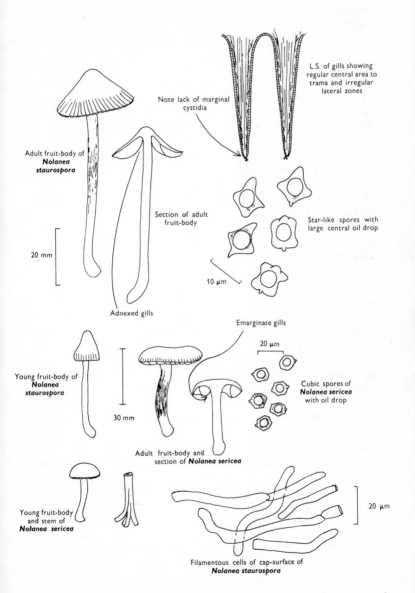

Adult fruit-body of **Nolanea staurospora**

Note lack of marginal cystidia

L.S. of gills showing regular central area to trama and irregular lateral zones

Section of adult fruit-body

Star-like spores with large central oil drop

20 mm

10 μm

Adnexed gills

Emarginate gills

20 μm

Young fruit-body of **Nolanea staurospora**

Cubic spores of **Nolanea sericea** with oil drop

30 mm

Adult fruit-body and section of **Nolanea sericea**

Young fruit-body and stem of **Nolanea sericea**

20 μm

Filamentous cells of cap-surface of **Nolanea staurospora**

except for the strong mealy smell; however, it is recognised immediately by the spore-shape, in fact stauro—means a cross and spora—
spore!

Because of the flattened cap and gill-shape *N. sericea* (Mérat)
P. D. Orton was first placed in *Entoloma*, but for a long time it was
one of the smallest members of that genus. The European species
of *Nolanea* have recently been critically analysed, and now that
closely related species to the silky *Nolanea* have been found, it
appears better placed in *Nolanea* although it is still found under
Entoloma in many books. The Latin word 'sericeum' means silky
and refers to the silky cap and stem of this fungus which is a very
noticeable feature when the fungus is collected in the dry state.
The common name which has been given to this fungus also refers
to the silky nature of the fruit-body.

Illustrations: N. staurospora—LH 181; ND 31^2; WD 52^2. *N. sericea*—
LH 181; WD 52^5.

Panaeolus foenisecii (Fries) Schroeter   Brown hay-cap
Cap: width 12–28 mm. *Stem:* width 3–6 mm; length 40–60 mm.
Description:

Cap: semiglobate to convex and hardly expanding even with age,
    smooth, expallent, dull cinnamon-brown or dark tan-colour, be-
    coming clay-colour or pale cinnamon-colour from centre outwards
    on drying and so sometimes appearing as if it is zoned.

Stem: slender, fragile, smooth and pale cinnamon-brown, except at
    apex where it is dotted with white; it is usually more brownish below.

Gills: adnate, crowded, pale brown and mottled, but becoming more
    uniformly umber-brown except for whitish margin.

Flesh: whitish or pale cinnamon colour.

Spore-print: purple-brown.

Spores: long, lemon-shaped under the microscope, dull brown,
    warted all over but for the distinct germ-pore; 12–15 × 7–8 μm in
    size.

Marginal cystidia: variable spindle-shaped with flexuous neck and sub-
    capitate apex, about 5–6 μm wide.

Facial cystidia: absent.

Habitat & Distribution: Common amongst short grass on lawns, in
    pastures, on grass-verges, etc., from May until October.

Plate 39. Fleshy fungi: Spores purple-brown and borne on gills

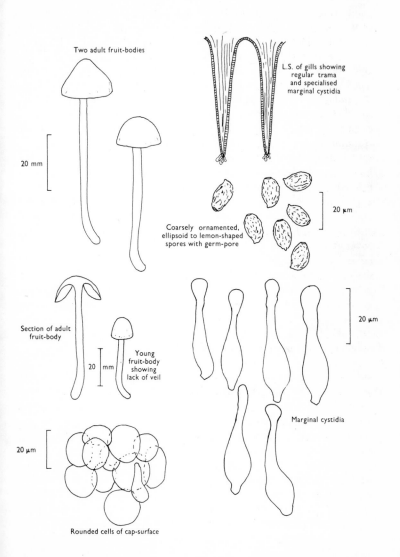

Two adult fruit-bodies

20 mm

L.S. of gills showing
regular trama
and specialised
marginal cystidia

20 µm

Coarsely ornamented,
ellipsoid to lemon-shaped
spores with germ-pore

20 µm

Section of adult
fruit-body

20 mm

Young
fruit-body
showing
lack of veil

Marginal cystidia

20 µm

Rounded cells of cap-surface

General Information: P. *foenisecii* is recognised under the microscope by the ornamented spores; this character was used to separate this fungus in the new genus *Panaeolina*. However, although the spore-print is not exactly black the stature, mottled gills and anatomy conform closely with *Panaeolus sphinctrinus* (Fries) Quélet and P. *rickenii* Hora (see p. 210 and below respectively). The same fungus has been placed in *Psilocybe* (see p. 114), but it has little in common with members of that genus. The word 'foenisecii' means hay-harvest, reflecting the habitat of growing in fields. This fungus is variable in colour depending on its state of turgidity; it can be easily confused with other species of *Panaeolus* when moist and with certain species of *Conocybe* when dry. P. *rickenii* is an equally common agaric growing on similar or slightly less base-rich soil-types. It has a distinctly bell-shaped reddish brown cap with a pale incurved margin which in wet weather is, like the entire stem, beaded with droplets of liquid. This gives the fungus a glistening appearance when seen fresh and as it dries these droplets are lost and the cap becomes slightly zoned. The stem is pale reddish-brown with a strong frosted appearance because of the minute hairs which cover it. I have no doubt that the classification of these fungi will be assisted by careful analysis of the shapes of the hairs found in the different species.

Illustrations: Panaeolina foenisecii—LH 145; WD 78[4]. *Panaeolus rickenii*—LH 145.

(f) Agarics of wasteland and hedgerows

Coprinus comatus (Fries) S. F. Gray   Lawyer's wig
Cap: width 30–60 mm; height 80–200 mm. *Stem:* width 10–20 mm; length 80–250 mm.
Description: Plate 40.
Cap: at first cylindrical or oval then bell-shaped, fleshy, fragile, white and covered with woolly, whitish, shaggy scales which have brown tips; the centre of the cap is smooth and yellow to ochraceous whilst the margin becomes striate and lilaceous and finally black as the tissue liquefies (autodigests) and the margin rolls up to expose new areas of spore-bearing tissue.
Stem: tall, white, smooth and tapered towards the apex, with a white

Plate 40. Fleshy fungi becoming reduced to an inky mass:
Spores black and borne on gills

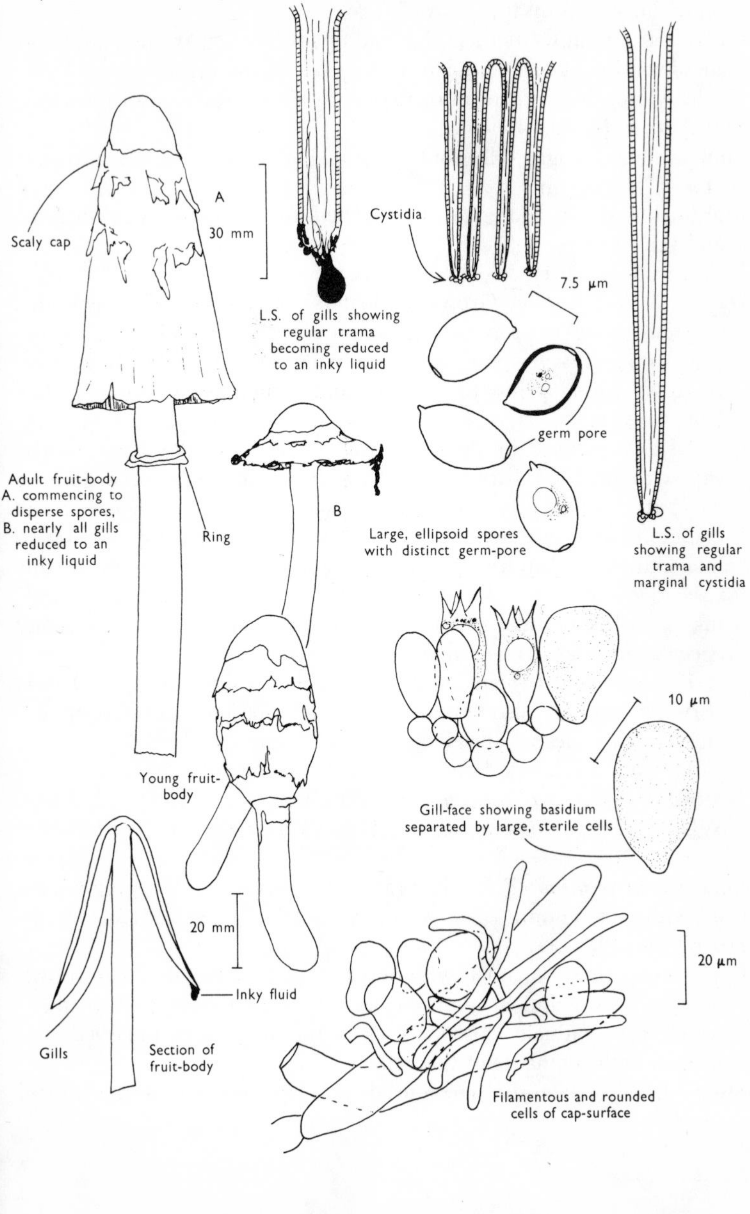

Scaly cap

A

30 mm

L.S. of gills showing
regular trama
becoming reduced
to an inky liquid

Cystidia

7.5 μm

germ pore

Large, ellipsoid spores
with distinct germ-pore

L.S. of gills
showing regular
trama and
marginal cystidia

Adult fruit-body
A. commencing to
disperse spores,
B. nearly all gills
reduced to an
inky liquid

Ring

B

Young fruit-
body

10 μm

Gill-face showing basidium
separated by large, sterile cells

20 mm

Inky fluid

20 μm

Gills

Section of
fruit-body

Filamentous and rounded
cells of cap-surface

ring which can easily move up and down the stem with handling, and which soon disappears with age.

Gills: free at first, white then pink and finally black, becoming gradually dissolved into a black fluid from the base of the cap upwards.

Flesh: white, thin, except immediately in the central area of the cap.

Spore-print: blackish-purple.

Spores: long, elongate-ellipsoid, large and about 13 × 5-8 μm in size, (12-15 × 7-9 μm).

Marginal cystidia: elongate club-shaped to balloon-shaped, hyaline and thin-walled.

Facial cystidia: absent.

Habitat & Distribution: Grows in clusters on rich ground, in gardens, on sides of newly prepared roads and central reservations of motor-ways, on path-sides, in cultivated fields and on rubbish dumps; it grows from spring to autumn and sometimes occurs in huge troops.

General Information: Easily recognised by its size, the shape of the cap with its scaly surface and from its resemblance to a 'judge's wig'; it is frequently called the 'lawyer's wig' and whereas some common names are not very descriptive and one has to use a lot of imagination to conjure up what the common name implies, in this case it is not so. It is also known as the 'shaggy cap' or 'shaggy ink-cap'. Ink or inky cap is, however, a common name for many species of the genus *Coprinus* (see p. 211-4).

The unrelated *Lyophyllum decastes* (Fries) Singer and *L. connatum* (Fries) Singer are also common fungi growing on roadsides, on soil and compost-heaps. They too break through embankments, soil, paths, etc., producing large craters and mounds of debris.

Illustrations: Coprinus comatus—F 34[b]; Hvass 172; LH 137; NB 35[5]; WD 82[2]. *Lyophyllum decastes*—LH 81; WD 14[2].

Lacrymaria velutina (Fries) Konrad & Maublanc   Weeping widow
Cap: width 45-90 mm. *Stem:* width 8-14 mm; length 50-125 mm.
Description: Plate 41.

Cap: convex then expanded with obtuse central umbo, dull clay-brown or date-brown and at first covered with flattened, woolly fibrils which are gradually lost with age; the margin is incurved and fringed with remnants of the veil.

Stem: fragile, pale dingy-coloured or clay-coloured at apex, dull

Plate 41. Fleshy fungi: Spores blackish and borne on gills

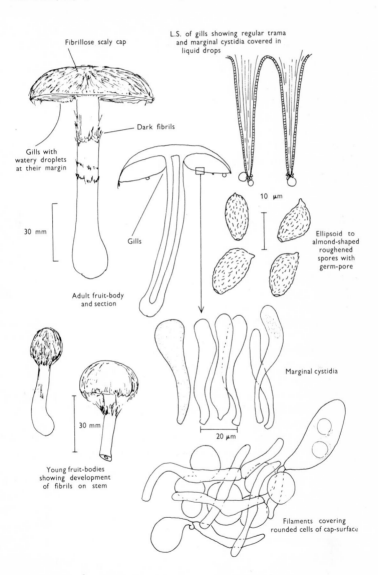

Fibrillose scaly cap

L.S. of gills showing regular trama and marginal cystidia covered in liquid drops

Dark fibrils

Gills with watery droplets at their margin

30 mm

10 μm

Gills

Ellipsoid to almond-shaped roughened spores with germ-pore

Adult fruit-body and section

Marginal cystidia

30 mm

20 μm

Young fruit-bodies showing development of fibrils on stem

Filaments covering rounded cells of cap-surface

brown below the ring-zone which consists of white fibrils; later in development these fibrils catch the spores and the stem becomes black and fibrillose-scaly, particularly below the ring-zone.

Gills: sinuate, crowded and very dark brown or almost black with distinct white margin which is covered in tiny beads of liquid in moist weather.

Flesh: pale buff.

Spore-print: almost black.

Spores: long, dark brown, lemon-shaped and warted with distinct and prominent germ-pore and 10–12 × 6–7 μm in size.

Marginal cystidia: club-shaped or with a distinctly rounded head.

Facial cystidia: absent.

Habitat & *Distribution:* Common on the ground near newly built houses, on roadsides, tips and paths in woods, either solitary or in groups; it is also found in pastures.

General Information: The fibrillose scaly cap and stem and the almost black gills which frequently have liquid droplets at their edge separate this species from all other agarics and microscopically it can be easily recognised by the warted spores. 'Velutina' means velvety and refers to the texture of the cap-surface, of the young fruit-body. The genus name *Lacrymaria* refers to this peculiar, but certainly not unique, phenomenon, of exuding liquid from cells on the gill-edge. This has been compared with weeping and 'lacrymans' means weeping; the common name reflects this also—weeping widow (cf. p. 154).

This fungus has had a chequered history, for it is also known in some books as *Hypholoma lacrymabunda* (again meaning weeping) or *H. velutina*; the anatomy of the fungus, however, is quite different to *Hypholoma* (e.g. *H. fasciculare* p. 64). More recently it has found a place in *Psathyrella*, but it seems unsatisfactorily placed there because of the warty spores, black spore-print and fibrillose cap-surface; it warrants a separate genus, i.e., *Lacrymaria. L. pyrotricha* (Fries) Konrad & Maublanc is the only other British species of this genus but it has a bright orange cap colour; it is rare.

Illustrations: Hvass 180; LH 141; WD 86[3].

Lepista nuda (Fries) Cooke   Wood blewits
Cap: width 70–100 mm. *Stem:* width 10–15 mm; length 70–100 mm.
Description: Plate 42.

Cap: rounded then flattened or slightly depressed in the centre, smooth, bluish lilac, or violaceous when young but gradually with age becoming reddish-brown, with or without a flush of wine colour.

Stem: similarly coloured to the cap, equal, fleshy, elastic, fibrillose and streaky.

Gills: adnate with or without a decurrent tooth, crowded, lilac and easily separable from the cap-tissue by the fingers.

Flesh: bluish violaceous, but drying out dirty buff in the base of the stem.

Spore-print: flesh-coloured.

Spores: medium-sized, ellipsoid appearing smooth but very minutely roughened under the microscope, although it is very difficult to see except with a good instrument (6–8 × 4–5 μm in size).

Marginal and facial cystidia: absent.

Habitat & *Distribution:* Widespread in troops or small groups in copses and under hedgerows and not uncommon in flower-beds in gardens in late autumn and early winter especially on compost heaps and in rhubarb patches which have been mulched with piles of moribund leaves.

General Information: This fungus was originally placed in *Tricholoma*, but due to differences in anatomy and the distinctly coloured and ornamented spores it has been placed along with 'common blewits' *T. personatum* (Fries) Kummer (or better *L. saeva* (Fries) P. D. Orton), in the genus *Lepista*. This genus which is also called *Rhodo-paxillus*, again referring to the pinkish spore-print, is not found in many of the easily obtainable books. One should look for the fungus under *Tricholoma*, from which it can be separated easily by the beautiful colour.

   Both the 'wood blewits' and 'common blewits' have been regularly sold in markets in England within the last fifty years. They are edible and considered of high quality. In their fresh state they are easily recognised, but as they age they become browned and so resemble many other less desirable fungi.

Illustrations: F 17[d]; Hvass 49; LH 91; NB 125[2]; WD 12[3] (a bit too pastel).

Plate 42. Fleshy fungi: Spores pale pinkish and borne on gills

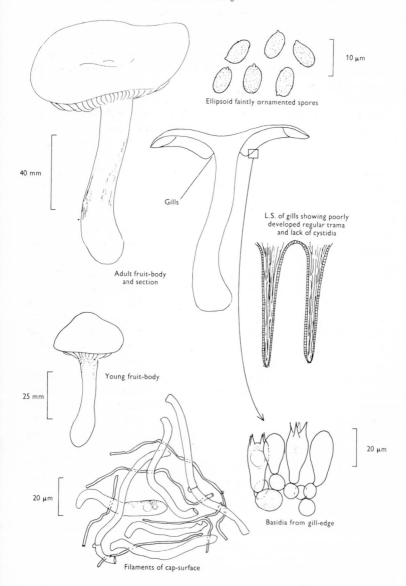

10 μm

Ellipsoid faintly ornamented spores

40 mm

Gills

Adult fruit-body
and section

L.S. of gills showing poorly
developed regular trama
and lack of cystidia

Young fruit-body

25 mm

20 μm

20 μm

Basidia from gill-edge

Filaments of cap-surface

Agaricus bisporus (J. Lange) Pilát   Common mushroom

Cap: width 40–100 mm. *Stem:* width 15–25 mm; length 50–75 mm.

Description: Plate 43.

Cap: rounded gradually expanding to become plane, whitish with numerous brown radiating fibrils and with the margin irregular because of fragments from the ring which are left there after expansion of the cap.

Stem: short, cylindrical, smooth, bruising reddish-brown when handled and with a narrow ring which soon collapses and disappears.

Gills: free, pink at first then purple-brown, narrow and crowded.

Flesh: solid, thick, firm and slowly flushing brownish on cutting.

Spore-print: purple-brown.

Spores: medium-sized, broadly ellipsoid, purple-brown under the microscope, less than 10 μm long, (6–8 × 5–6 μm).

Marginal cystidia: club-shaped, 10–12 μm at apex.

Facial cystidia: absent.

Basidia: 2-spored.

Habitat & *Distribution:* Frequent on manure heaps, straw heaps, on road scrapings and around garden plants.

General Information: This fungus is recognised by the dark fibrils on the cap, the 2-spored basidia easily seen with the low power of a microscope, and the pink gills when young. Much confusion has existed over this fungus and its nearest relatives. It is similar to the 'Cultivated mushroom', *A. hortensis* (Cooke) Pilát, which is offered for sale in shops. However, it differs in several minor details and it may be that *A. bisporus* is the fungus from which the cultivated mushroom developed, very probably unconsciously by man, but the history of the cultivated mushroom is very obscure. The cultivated mushroom when bought in British shops is white but in the United States two varieties are sold, one with the brownish fibrils predominating and a snow-white one where the fibrils do not darken; the former is frequently found in Europe. The white form is sometimes found in gardens where spent-mushroom spawn is used as mulching around fruit-trees but it has a rounder cap than *A. bisporus.* The cultivated mushroom accounts for an annual income of £14 million in the British Isles.

Illustrations: A. hortensis—LH 133 (as the forma *albida*); NB 31[7]; WD 71[1]. *A. bisporus*—Hvass 161; LH 133.

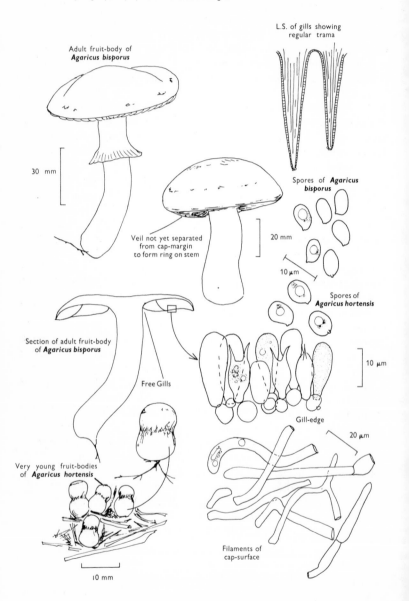

Plate 43. Fleshy fungi: Spores purple-brown and borne on gills

L.S. of gills showing regular trama

Adult fruit-body of *Agaricus bisporus*

30 mm

Spores of *Agaricus bisporus*

Veil not yet separated from cap-margin to form ring on stem

20 mm

10 μm

Spores of *Agaricus hortensis*

Section of adult fruit-body of *Agaricus bisporus*

Free Gills

10 μm

Gill-edge

Very young fruit-bodies of *Agaricus hortensis*

20 μm

Filaments of cap-surface

10 mm

B. BRACKET-FUNGI AND THEIR RELATIVES

Key to major genera

A group of fungi which includes the bracket fungi, hedgehog fungi, fairy-clubs and their relatives; in the majority of species the margin continues to grow through the favourable part of the season and so often envelopes leaves, grass, etc.

1. Spore-bearing layer (hymenium) quite smooth, spread over veins or shallow pores; fruit-body top-shaped, fan-shaped or club-shaped, or spread over the substrate (resupinate) .. 2
   Spore-bearing layer lining the inner surface of tubes or borne on warts or spines 17

2. Fruit-body club-shaped, coral-shaped or distinctly funnel-shaped, fan-like or resembling an agaric 3
   Fruit-body resupinate or with poorly developed cap 11

3. Fruit-body coral-like or club-shaped with clubs grouped or branched 4
   Fruit-body resembling an agaric or funnel-shaped to fan-shaped 9

4. Fruit-body large, branched with flattened and curled lobes and so resembling a cauliflower .. .. .. *Sparassis*
   Fruit-body of single or grouped clubs or if branched then not resembling a cauliflower, the lobes being cylindrical or only slightly flattened and hardly bent 5

5. Fruit-body small arising from a seed-like structure or growing attached to dead herbaceous plant remains 6
   Fruit-body medium to large, simple or branched and usually growing on the ground; one large species grows on wood .. 7

6. Fruit-body arising from a seed-like body embedded in the plant-tissue or found loose in the soil .. .. .. *Typhula*
   Fruit-body on dead plant-remains but seed-like structure absent .. .. *Pistillaria*

7. Fruit-body much branched; spores ornamented (see also
   Thelephora below)   ..   ..   *Ramaria*
   Fruit-body simple or if with well-developed branches then
   spores smooth   8
8. Fruit-body branched irregularly with many to few branches,
   grey, white or drab-coloured; spores large, subglobose and
   smooth   ..   ..   *Clavulina*
   Fruit-body club-shaped or if branched then brightly coloured
   and spores not large and subglobose
                      Clavaria, Clavulinopsis & *Clavariadelphus*
9. Fruit-body resembling an agaric with spores borne on fold-
   like, often forked and shallow ridges and veins, and often
   brightly coloured
              Cantharellus (compare carefully with *Craterellus* below)
   Fruit-body funnel-shaped or fan-shaped   10
10. Fruit-body often drab colour or greyed with smooth or
    slightly veined outer surface   ..   ..   *Craterellus*
    Fruit-body wrinkled, irregular or smooth and powdery,
    lilaceous to chocolate-brown in colour   ..   *Thelephora*
11. Fruit-body sessile or resupinate and fleshy; spores borne on
    veins united to form shallow pores   12
    Fruit-body resupinate or bracket-like, and spore-surface
    veined or rugulose but lacking distinct pores   13
12. Spores colourless   ..   ..   *Merulius*
    Spores brown   ..   ..   *Serpula*
13. Spore-bearing layer containing long, brown spines   *Hymenochaete*
    Fruit-body lacking spines although often having encrusted
    sterile cells   14
14. Surface of fruit-body more or less radiately veined   ..   *Phlebia*
    Surface of fruit-body not radiately veined   15
15. Spores brown   ..   ..   *Coniophora*
    Spores colourless   16
16. Flesh distinctly formed and fruit-body with or without a well
    formed cap   ..   ..   ..   *Stereum* & related genera
    Flesh poorly differentiated and fruit-body lacking a cap
    members of the *Corticiaceae* (including *Peniophora* & *Hyphodontia*
                                                           p. 176)
17. Spores borne on teeth or spines   18
    Spore-bearing layer lining tubes or elongate pores   22

18. Fruit-body with central stem; agaric-like but not attached to cones   19

  Fruit-body encrusting or bracket-like, or with lateral stem if resembling an agaric   20

19. Fruit-body fleshy   .. .. *Hydnum* and related genera

  Fruit-body rubbery or tough   .. *Hydnellum* and related genera

20. Fruit-body growing attached to cones and cap with lateral stem   .. .. *Auriscalpium*

  Fruit-body not on cones and distinct stem lacking   21

21. Spores borne on a series of radially arranged knotches resembling gills ..   .. *Lentinellus*

  Spores borne on a resupinate layer of spines

  Mycoacia and related genera

22. Tubes free one from another   .. .. *Fistulina*

  Tubes united to form a distinct tissue 23

23. Fruit-body perennial and exhibiting more than one layer of tubes   24

  Fruit-body annual although the fruit-body can persist in a dried depauperate form for several months   27

24. Spores brown   25

  Spores colourless   26

25. Large, brown, sterile cells present in the tubes; spores simple

  Phellinus & *Cryptoderma*

  Brown, sterile cells absent from tubes; spores complex *Ganoderma*

26. Large woody fruit-body with crust-like top   .. .. *Fomes*

  Medium sized to small, fleshy-tough fruit-body with downy or crust-like top   .. *Oxyporus*, *Fomitopsis* & *Heterobasidion*

27. Spores borne in labyrinth-like or elongate pores, or cap either poorly developed or absent, and only resupinate pore-surface present   28

  Spores borne in distinct pores on well-developed woody fruit-bodies   31

28. Spores borne in labyrinth-like pores   *Daedalea* & *Daedaleopsis*

  Spores borne in elongate pores like very thick gills, or fruit-body completely resupinate   29

29. Spore-layer lining elongate pores

  Lenzites (white) & *Gloeophyllum* (brown)

  Spore-layer consisting of a resupinate pore-layer   30

30. Pore-layer totally resupinate; flesh very poorly developed
                                    Fibuloporia and related genera
    Fruit-body resupinate or developing ill-formed caps at the
    margin; flesh well-developed and quite tough
                        Datronia, Gloeoporus & *Bjerkandera*
31. Fruit-body with a distinct stem   32
    Fruit-body sessile or with a poorly developed stem, or if
    merely with a basal swelling then pores bruising   33
32. Pores dark-coloured but spores pale-coloured in mass
                            Coltricia (also see *Phaeolus* below)
    Pores white or creamy, foot often darkened or black, and
    spores hyaline   ..   .. *Polyporus*
33. Pores brightly coloured, red, lilaceous or orange to apricot-
    colour   34
    Pores never as brightly coloured, cream, white, grey or in
    some shade of brown   35
34. Pores red to orange-red   ..   ..   ..   *Pycnoporus*
    Pores lilac to violaceous, or lilaceous orange to apricot colour
        Hapalopilus (orange-apricot) & *Hirschioporus* (lilaceous)
35. Pore-surface brown or dark grey and spores often colourless   36
    Pore-surface white or creamy, or yellow; spores hyaline   .. 38
36. Pore-surface firm and grey   ..   ..   ..   *Bjerkandera*
    Pore-surface greenish yellow, bruising brown or yellow-
    brown and darkening with age   37
37. Fruit-body lacking a stem, rust-brown, breaking easily,
    cheesy and with silky sheen   ..   ..   ..   *Inonotus*
    Fruit-body with a broad basal hump, fibrillose spongy with
    yellow margin to cap   ..   .. *Phaeolus*
38. Tubes forming a layer quite distinct from the flesh; fruit-body
    fleshy and tough   39
    Tubes not forming a layer distinct from the flesh; fruit-body
    woody or corky 43
39. Pore-surface bright yellow; upper surface yellow or orange
                                                   Laetiporus
    Pore-surface white   40
40. Fruit-body medium to large, shell-shaped, whitish brown or
    silvery grey on top; on birch   ..   .. *Piptoporus*
    Fruit-body often frond-like, infrequently shell-shaped and
    if on birch then small   41

41. Fruit-body fan- or frond-shaped, composed of innumerable more or less complete caps joined together at their base or to half-way .. .. *Grifola* & *Meripilus*
Fruit-body neither fan-shaped nor frond-shaped and compound 42

42. Fruit-body wholly pale-coloured white, cream, ivory, etc.
Tyromyces
Fruit-body except pores usually some shade of brown *Polyporus*

43. Cap thick, corky or woody and pores medium or large
Trametes & *Pseudotrametes*
Cap thin but leathery and pores small .. .. *Coriolus*

(i) Pored and toothed fungi

(a) Colonisers of tree trunks, stumps and branches

Polyporus squamosus Fries   Scaly polypore
Cap: 100–300 mm. *Stem:* width 25–50 mm; length 25–75 mm.
Description:

Cap: fan-shaped or semicircular, spreading horizontally with age, ochre-yellow or straw-coloured with dark brown, flattened scales in concentric zones which are much more dense at the centre.

Stem: short, stout, white at apex and netted with pale creamy buff about middle, but dark brown or black towards the base and attached to the side of the cap.

Tubes: whitish to yellowish and decurrent with large, angular, irregularly fringed, whitish or cream-coloured pores.

Flesh: with strong, not very pleasant smell, cream-coloured or white.

Spore-print: white.

Spores: long, oblong or elongate ellipsoid, hyaline under the microscope (10–15 × 4–5 μm) and not blueing in solutions containing iodine.

Habitat & *Distribution:* An easily recognisable fungus growing on stumps and old living trees, especially of sycamore and elm where it often forms tiers of caps from late spring until autumn; however, they decompose rapidly and almost completely disappear by the next year when new fruit-bodies may appear in the same place, a phenomenon which may take place for several consecutive seasons.

General Information. The genus *Polyporus* is in most text-books, a big and unwieldy genus joining together all fleshy, annual fungi possessing tubes; even the boleti (see p. 32) have been included! Many of these species are now considered less closely related one to another than previously thought. Boleti differ from polypores, however, in their less tough and distinctly putrescent fruit-body, and in the fact that the margin of the cap extends but does not continue to grow during the life-cycle; the margin of the polypore fruit-body is active and may burst into growth again when favourable weather conditions occur. The 'Scaly polypore' has a flesh which consists of two types of hyphae: (i) hyphae of unlimited growth with abundant protoplasmic contents which stain easily and which collapse on drying; and (ii) thick-walled, strengthening hyphae which bind the thin walled hyphae together. *Laetiporus sulphureus* (Fries) Murrill 'Sulphur polypore' has a single type of hyphae in

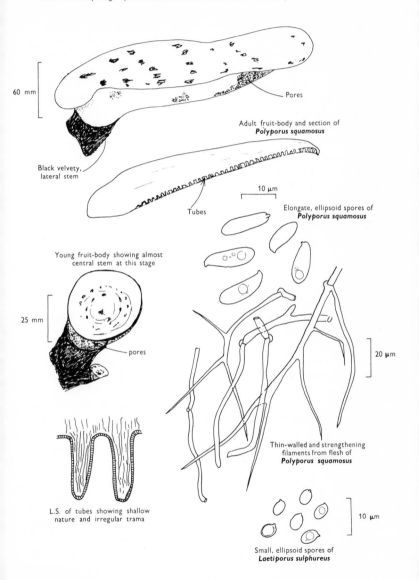

Plate 44. Woody fungi: Spores white and borne within tubes — fruit-body annual

60 mm

Pores

Adult fruit-body and section of
Polyporus squamosus

Black velvety,
lateral stem

Tubes

10 μm

Elongate, ellipsoid spores of
Polyporus squamosus

Young fruit-body showing almost
central stem at this stage

25 mm

pores

20 μm

Thin-walled and strengthening
filaments from flesh of
Polyporus squamosus

L.S. of tubes showing shallow
nature and irregular trama

10 μm

Small, ellipsoid spores of
Laetiporus sulphureus

the tubes, i.e. thin walled generative, and only a few binding hyphae in the flesh. It has an orange cap with a rather thick, sulphur or chrome-yellow margin, sulphur-yellow tubes and pores and yellow, then pale buff, flesh. The spore-print is white and the spores hyaline, pip-shaped and medium sized, (5–7 × 4–5 μm).

Illustrations: P. squamosus—F 43[b]; Hvass 267; LH 75; NB 129[1]; WD 94[1]. *L. sulphureus*—Hvass 268; LH 73; NB 129[3]; WD 94[2].

Some common annual polypores

Piptoporus betulinus (Fries) Karsten   Birch polypore

Cap: 75–200 mm, kidney-shaped or hoof-shaped, smooth, covered by a thin, separable and greyish silvery or pale brownish skin; cap-margin thick, incurved and projects beyond the tubes.

Stem: rudimentary, simply a small hump below which the fungus develops.

Tubes, pores and spore-print: white.

Spores: sausage-shaped, and thin-walled hyaline under the microscope and very narrow, (5–6 × 1–2 μm). It grows on birch throughout the country where it causes a sap wood-rot which finally converts the inner timber to a red-brown friable mass. The flesh, which contains thickened binding hyphae, is used for mounting insects and for sharpening knives, hence the common name 'Razor-strop fungus'.

Illustrations: Hvass 269; LH 67; NB 117[4]; WD 93[3].

Inonotus hispidus (Fries) Karsten   Shaggy polypore

Cap: 100–250 mm, kidney-shaped, yellow-brown to rust-brown, but finally almost black, at first covered with shaggy hairs, but these tend to mat together with age.

Stem: absent.

Tubes and flesh: rust-colour; pores at first yellow, but finally red-brown.

Spore-print: yellow-brown.

Spores: medium sized (8–9 × 7–8 μm) and globose under the microscope. It grows on various broad leaved trees, especially ash where it causes a spongy, white heart-wood rot. The flesh contains hyphae with thick, brown walls.

Illustrations: LH 63; WD 96[1].

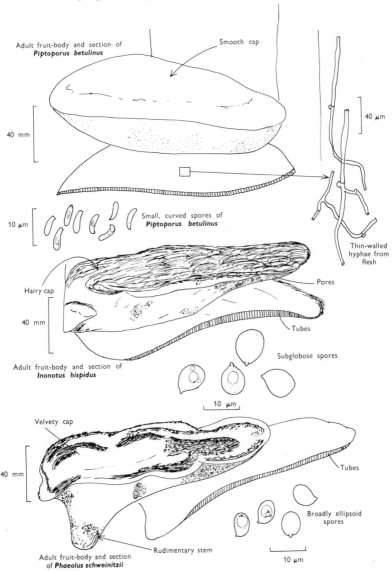

Plate 45. Woody fungi — annual polypores

Adult fruit-body and section of
Piptoporus betulinus

Smooth cap

40 mm

40 µm

10 µm — Small, curved spores of
Piptoporus betulinus

Thin-walled
hyphae from
flesh

Hairy cap

Pores

40 mm

Tubes

Adult fruit-body and section of
Inonotus hispidus

Subglobose spores

10 µm

Velvety cap

40 mm

Tubes

Rudimentary stem

Broadly ellipsoid
spores

Adult fruit-body and section
of **Phaeolus schweinitzii**

10 µm

Phaeolus schweinitzii (Fries) Patouillard

Cap: 100–300 mm, bracket-shaped or tub-shaped, dark brown with a knobbly, velvety, roughened and grooved surface; margin at first golden yellow.

Stem: absent or short, thick and brown.

Tubes and pores: greenish yellow.

Flesh: deep rust-brown.

Spore-print: greenish yellow.

Spores: medium sized, greenish under the microscope, ellipsoid and about 8 × 4 μm in size, (7–8 × 3–4 μm). This fungus is found on conifers or near conifer stumps where it is attached to the roots; it causes a brown cubical heart-wood rot; the flesh of the fruit-body is composed of only one type of hyphae.

Illustrations: LH 67; NB 111[3]; WD 95[1].

Meripilus giganteus (Fries) Karsten   Giant polypore

Cap: 75–100 mm, or even up to 200 mm wide, grouped and forming a tuft of caps up to 750 mm across. The individual caps are fan-shaped, pliable, rather thin and yellow-brown to snuff-brown with their margins wavy and cream colour or yellowish.

Stem: replaced by a united mass of caps.

Tubes, pores and flesh: white and very soft, but becoming black on bruising.

Spores: small, pip-shaped, hyaline under the microscope and 5–6 × 4–5 μm. This fungus is a common sight forming masses at the base of broad-leaved trees; it is common on beech. It is a soft, fibrous polypore as a result of the lack in the flesh of thick-walled specialised hyphae.

Illustrations: Hvass 277; LH 73; NB 129[4]; WD 93[1].

The spores of all the annual polypores described above do not blue when placed in solutions containing iodine.

Coriolus versicolor (Fries) Quélet Many zoned polypore
Cap: 25–60 mm. *Stem:* absent.
Description: Plate 46.
Cap: semi-circular, flattened, thin, tough and flexible when fresh with
the surface velvety and marked with smoother, paler concentric
zones giving a pattern of yellow-brown, grey or darker greenish
grey zones; the margin is thin and is the palest of the zones and
may be wavy or lobed.
Tubes: white with small, round and rough-edged to angular white or
cream-coloured pores which become yellowish with age.
Flesh: white, tough and continuous with the tube tissue and so not
allowing one to detect any difference between the tissues.
Spore-print: white.
Spores: medium sized, oblong and hyaline under the microscope, and
6–8 $\times$ 2–3 μm; not blueing in solutions containing iodine.
Habitat & *Distribution:* Very common on stumps, trunks and fallen
branches of various trees, especially beech; it is to be found throughout the year.
General Information: It is often associated with nodulose masses of
fungal tissue which are covered in small poroid areas and are very
confusing when found by the beginner; they are simply growth-
forms of *Coriolus versicolor*; such forms are frequently found on
old house-timbers exposed to the weather, particularly window
frames where it forms a distinct rot. Its flesh consists of thin-walled
hyphae and binding hyphae as in *Polyporus squamosus* as well as an
additional thick-walled type called skeletal hyphae. It would appear
that several polypores are capable of producing the amorphous
growths mentioned above, some of which contain hyphal fragments
called conidia.

The bands of colour on the cap of the 'many zoned polypore' are
retained after drying and from a group of fruit-bodies the most
attractively zoned can be selected, mounted on small pieces of wood
or cardboard and fitted at the back with a pin. Such preparations
make very attractive brooches and have been used even by modern
designers to contrast with their fashion creations.

There are many pale tubed polypores growing on wood. *Daedalea
quercina* Fries 'Mazegill', grows on oak and has irregular maze-like
pores; *Lenzites betulina* (Fries) Fries, grows on birch, has tough
plates which resemble the gills of an agaric. *Datronia mollis* (Fries)

Donk forms thick spreading resupinate patches on beech, sometimes with irregular dark brown caps formed by the upturned margin. Several species of *Tyromyces* occur in Britain and are characterised by their white pores and tubes and the white or pale-coloured caps. *Bjerkandera adusta* (Fries) Karsten has a grey pore-surface and is also frequently found on beech.

Illustrations: F 44a; LH 69; NB 117³; WD 51².

Ganoderma europaeum Steyaert   Common ganoderma

Cap: 100–350 mm. *Stem:* absent.

Description: Plate 47.

Cap: bracket-shaped, rather flat at margin but humpy and irregular about the middle, frequently concentrically zoned, smooth and only slightly shiny; its margin is whitish or pale greyish.

Tubes: red-brown or cinnamon-brown, obscurely layered and with small, white pores flushed with pale cinnamon-brown, but deep red-brown when rubbed or with age.

Flesh: with a fragrant smell, deep red brown and felty-fibrous.

Spore-print: dark cinnamon-brown.

Spores: long, oval with truncate apex, smooth, but reticulate on the inner surface of the inner wall giving the spores a patterned appearance when seen under the microscope; 10–11 × 6–7 µm in size.

Habitat & Distribution: This fungus is common on various trees, especially beech and can be found throughout the year.

General Information: This common *Ganoderma* is perennial and distinguished from other polypore groups by the complex spores. *G. applanatum* (Fries) Karsten is closely related, but differs in the thinner fruit-body with a thin margin, and the pale cinnamon-brown flesh; the flesh of both species contains thick-walled binding and strengthening hyphae as well as the generative hyphae.

So sensitive are the pores to bruising that if a drawing or writing is executed on the lower surface with a pin, needle or similar sharp instrument and the fungus dried, the red-brown lines produced are

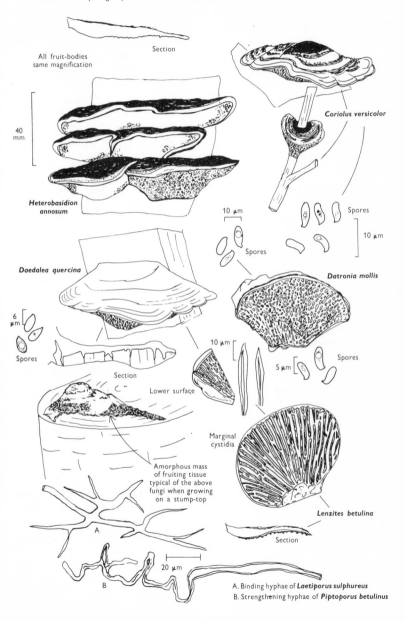

Plate 46. Woody fungi: Spores white and borne within tubes or on thickened plates

Section

All fruit-bodies
same magnification

Coriolus versicolor

40
mm

*Heterobasidion
annosum*

10 μm

Spores

Spores

10 μm

Daedalea quercina

Datronia mollis

6
μm

Spores

10 μm

Spores

5 μm

Section

Lower surface

Marginal
cystidia

Amorphous mass
of fruiting tissue
typical of the above
fungi when growing
on a stump-top

Lenzites betulina

A

Section

B

20 μm

A. Binding hyphae of *Laetiporus sulphureus*
B. Strengthening hyphae of *Piptoporus betulinus*

retained and the pattern preserved. Several fungus paintings prepared in this way were made in the early part of the century, many beautiful ones having originated in the eastern part of North America.

Fomes fomentarius whose important characters are described below has frequently been confused with *Ganoderma europaeum*. It is common growing on birch in Scotland, but is less frequent south of Perth, and then grows probably more frequently on beech which is similar to the pattern found on the continent of Europe. However, it has grown in former periods in England on birch, for it was found commonly amongst birch timbers in an excavation of an early Mesolithic lake side village near Scarborough, Yorkshire.

Illustrations: NB 125[3]; WD 160[2].

Some perennial polypores. Plate 48.

Fomes fomentarius (Fries) Kickx   Tinder fungus

Cap: 90–300 mm, hoof-shaped, thick, broadly attached to the substrate, zoned with yellow-brown and shades of grey; its margin is blunt and fawn or pale brownish.

Tubes: layered, cinnamon-brown with pale cinnamon pores with a whitish bloom.

Flesh: cinnamon-brown or buff and woolly.

Spore-print: white.

Spores: elongate, ellipsoid, very long, hyaline under the microscope, 15–18 × 5–6 μm, and not ornamented. The flesh contains both thick- and thin-walled hyphae. It grows on birch and less frequently on beech. The flesh has been used in dentistry, in manufacturing fancy articles, such as mats, and was the basis of the tinder used in flint-boxes.

Illustrations: LH 65; NB 117[1]; WD 100[1].

Phellinus igniarius (Fries) Quélet 'Willow Fomes', grows on willows and causes their heart-rot. It is a rust-brown, woody fungus with a hard crust and brown tubes and flesh. The spore-print is white and composed of small, spherical, hyaline spores, 5–6 μm in diameter. The flesh contains thin- and thick-walled hyphae.

Illustrations: LH 63; WD 99[3].

Plate 47. Woody fungi: Spores brown and borne within tubes — fruit-body perennial

Ganoderma europaeum

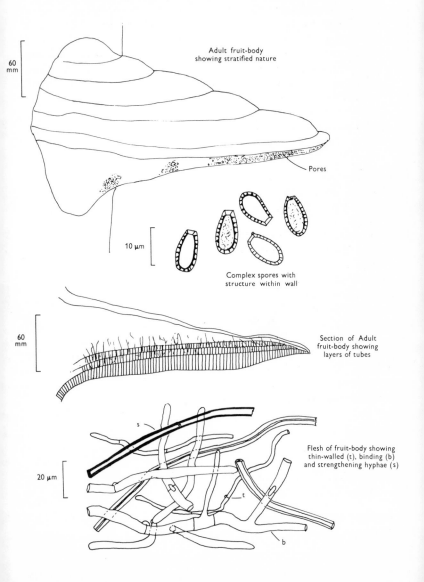

60 mm

Adult fruit-body
showing stratified nature

Pores

10 μm

Complex spores with
structure within wall

60 mm

Section of Adult
fruit-body showing
layers of tubes

20 μm

Flesh of fruit-body showing
thin-walled (t), binding (b)
and strengthening hyphae (s)

Oxyporus populinus (Fries) Donk, grows on various sorts of broad-leaved trees, particularly poplars and often becomes covered in mosses and algae. It has a pale buff or cream-coloured cap, white flesh, pores, tubes and spores.
Illustrations: LH 67.

Cryptoderma pini (Fries) Imaz, grows on conifers often several feet above the ground. It has a woody, deeply cracked upper surface, dark red-brown flesh, tubes and pores. Its spores are small, broadly ellipsoid and brown.

Heterobasidion annosum (Fries) Brefeld   Root fomes
Variable, sometimes possessing a cap, sometimes resupinate except for the upturned margin, flattened or shell-shaped, red-brown to blackish at the centre but pale at the margin, which when seen from below is always white. The tubes are in layers and like the pores, flesh and spore-print are white. The spores are broadly ellipsoid, small, smooth, hyaline and 4–5 × 4 μm. The flesh is fairly tough as it contains both generative hyphae and skeletal hyphae. It is frequent on the roots and lower parts of stems of many trees and shrubs causing a rapid heart-rot of conifers and extensive damage to young trees.
Illustrations: LH 67; NB 111[1]; WD 98[1].

The spores of all the perennial polypores described above do not blue when placed in solutions containing iodine.

Plate 48. Woody fungi: Spores borne within tubes — perennial polypores

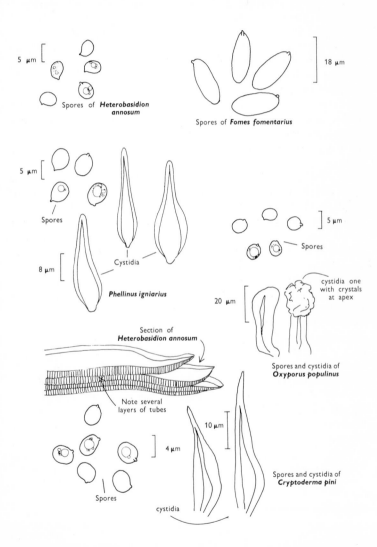

5 µm

Spores of **Heterobasidion annosum**

18 µm

Spores of **Fomes fomentarius**

5 µm

Spores

Cystidia

8 µm

Phellinus igniarius

5 µm

Spores

cystidia one with crystals at apex

20 µm

Spores and cystidia of **Oxyporus populinus**

Section of **Heterobasidion annosum**

Note several layers of tubes

10 µm

4 µm

Spores

cystidia

Spores and cystidia of **Cryptoderma pini**

Schizophyllum commune Fries   Split-gill fungus
Cap: 10–25 mm. *Stem:* width 2–4 mm; length 2–4 mm.
Description:
Cap: greyish fawn becoming whitish when dry, fan or kidney-shaped, often lobed and covered in close-set hairs and with incurved margin.
Stem: absent or the cap simply narrows into a stem-like bump.
Gills: replaced by a series of grey-brown plates which when dry appear as if to split longitudinally and their edges roll back.
Flesh: brownish but drying whitish.
Spore-print: white.
Spores: medium sized, oblong, hyaline under the microscope, not blueing in solutions containing iodine and 6–7 × 2–5 μm in size.
Facial and marginal cystidia: absent.
Habitat & Distribution: Grows on fallen branches, trunks, dead wood, etc.
General Information: Easily recognised by the 'gills' radiating from a point and becoming 'split' when dry. Specimens of *Schizophyllum* sealed by A. H. R. Buller in a tube in 1911 have been shown on remoistening to unroll their gills and shed variable spores, after 52½ years—probably a world record! The split-gill is a rather unique British fungus which appears to be much more closely related to the polypores than to the agarics—although it has for a long time been associated with the Oyster mushroom (p. 74). In fact, the splitting gills are two adjacent shallow dishes with spores produced on their inner surfaces. The cups separate on drying and therefore only superficially resemble gills splitting down the centre.

Another fungus which can also be associated with the idea of cups is *Fistulina hepatica* Fries 'the Beef-steak fungus'. This fungus is a polypore in the widest sense. It may grow up to 250 mm wide and is reddish-brown or liver-coloured with reddish tubes and pale flesh-coloured pores; the tubes although free are aggregated together and can be easily separated individually with the fingers. This fungus is edible although very strong in taste, it produces a serious decay of oaks.
Illustrations: S. commune—LH 105; NB 125[6]; WD 69[3]. F. hepatica—F 43[a] (lower figure); Hvass 278; LH 75; NB 129[2]; WD 101[4].

Plate 49. Woody fungi: Spores white and borne on split-'gills'

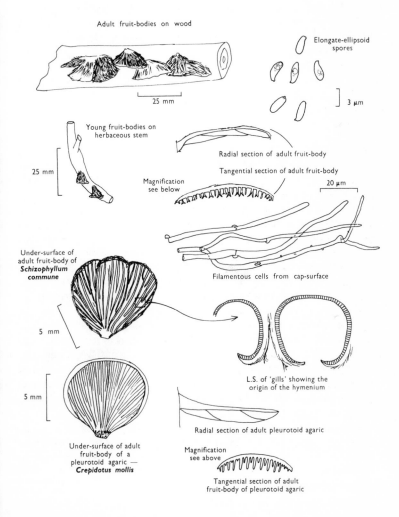

Adult fruit-bodies on wood

25 mm

Elongate-ellipsoid spores

3 µm

Young fruit-bodies on herbaceous stem

25 mm

Radial section of adult fruit-body

Tangential section of adult fruit-body

Magnification see below

20 µm

Under-surface of adult fruit-body of **Schizophyllum commune**

Filamentous cells from cap-surface

5 mm

L.S. of 'gills' showing the origin of the hymenium

5 mm

Under-surface of adult fruit-body of a pleurotoid agaric — **Crepidotus mollis**

Radial section of adult pleurotoid agaric

Magnification see above

Tangential section of adult fruit-body of pleurotoid agaric

(b) Destroyers of timber in buildings

Serpula lacrymans (Fries) Karsten   Dry-rot fungus
Description:

Fruit-body: usually widely spreading, but sometimes forming a distinct bracket with the upper surface silvery or smokey grey, flushed with lilac or rose or yellowish.

Stem: absent and replaced by a series of dirty white or greyish mycelial threads or strands which can be traced up to 100 mm over the substrate.

Flesh: thin, dirty yellowish and composed of only one type of hypha.

Spores: borne in shallow pores which are part of a complicated network of rust-brown folds and ridges.

Spore-print: rust-brown.

Spores: medium sized, golden yellow, thick-walled and broadly ellipsoid, and 8–10 × 5–6 µm in size.

Cystidia: absent.

Habitat & *Distribution:* On worked wood in buildings and less commonly in timber-yards. It can be found throughout the year.

General Information: This fungus forms fan-like structures and strands of mycelium which pass along beams and joists and through plaster. Where there is a bad case of dry-rot, the room or building will have an unpleasant musty smell and when actually growing the fungus exudes droplets of water on the mycelium and fruit-body, i.e. weeping, hence the name 'lacrymans'—weepy. It is a very important and destructive agent causing damage to floors and skirting boards, to joists and beams. It is a frequent pest of old houses and therefore of many of our cities. This fungus does not appear to have been found in the wild in Europe, but there is a record from the Himalayas. There are, however, very closely related species found on soil or wood-detritus. The Dry-rot fungus darkens the wood and produces a rot which makes the wood crack into small cubes or rectangular blocks.

This fungus was formerly placed in *Merulius*, but this genus should be retained for hyaline-spored fungi, e.g. *M. tremellosus* Fries, a species which grows even in winter on stumps of various trees in our woods.

Illustrations: LH 53; WD 103[3].

Plate 50. Dry-rot fungi — leathery and tough spores borne in shallow irregular pores

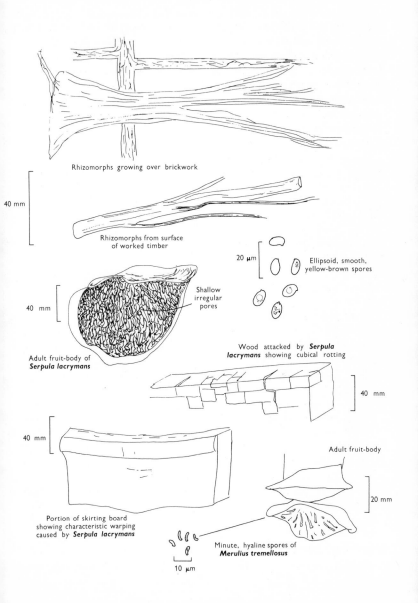

Rhizomorphs growing over brickwork

40 mm

Rhizomorphs from surface of worked timber

20 µm

Ellipsoid, smooth, yellow-brown spores

40 mm

Shallow irregular pores

Adult fruit-body of *Serpula lacrymans*

Wood attacked by *Serpula lacrymans* showing cubical rotting

40 mm

40 mm

Adult fruit-body

20 mm

Portion of skirting board showing characteristic warping caused by *Serpula lacrymans*

Minute, hyaline spores of *Merulius tremellosus*

10 µm

Coniophora puteana (Fries) Karsten   Cellar or Wet-rot fungus
Description:
Fruit-body: variable in size, resupinate, composed of one type of
   hypha only and with a sterile whitish cream or yellow margin.
Spore-bearing tissue: an irregularly wrinkled or humpy, yellowish
   surface which then becomes olive-green or bronze-colour.
Spore-print: olivaceous brown.
Spores: olive-brown under the microscope, smooth, ellipsoid, thick-
   walled and 12–14 × 8–9 μm in size.
Cystidia: absent.
Habitat & *Distribution:* This fungus causes wet-rot in houses, but may
   also be found on stumps and fallen trunks in woodland.
General Information: The fungus causes a discolouration of worked
   timber and induces longitudinal cracking with only a few lateral hair-
   like cracks unlike timber attacked by the dry-rot fungus (see p. 154).
Illustrations: WD 103[5].

Fibuloporia vaillantii (Fries) Bondarsev & Singer
Description:
Fruit-body: a resupinate layer of pores with cream-coloured or white
   sterile radiating margin.
Spore-bearing tissue: distributed within a series of small often shallow,
   white or ivory tubes.
Spore-print: white.
Spores: smooth, hyaline under the microscope, oblong 5–7 × 3–4 μm.
Cystidia: absent.
Habitat & *Distribution:* The dry-rot of houses, particularly in roof-
   systems.
General Information: Fibuloporia vaillantii is recognised by the white,
   resupinate pore-surface and fairly tough nature due to the presence
   of strengthening hyphae. Just as the genus *Polyporus* was found to
   be composed of several quite different elements (see pp. 140–44) and
   has since been split up into a number of different genera, the genus
   Poria has also been fragmented; one of the constituent genera is
   Fibuloporia. Amyloporia xantha (Fries) Bondarsev & Singer differs
   in having amyloid tissue and cystidia encrusted with crystals. The
   flesh contains both simple hyphae and thickened structural hyphae.
   It is yet another member of the large old unwieldy genus *Poria*
   and causes decay of worked wood, particularly the timbers of bench-
   ing and staging in greenhouses. *A. xantha* has a sulphur-yellow pore-
   surface and is rather cheesy when handled.

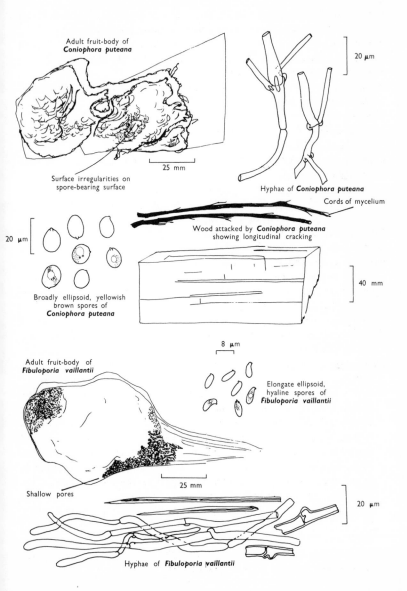

Plate 51. Wet and Dry-rot fungi — leathery and tough and spores borne within shallow pores or on an uneven surface

Adult fruit-body of
Coniophora puteana

20 μm

Surface irregularities on
spore-bearing surface

25 mm

Hyphae of *Coniophora puteana*

Cords of mycelium

20 μm

Wood attacked by *Coniophora puteana*
showing longitudinal cracking

Broadly ellipsoid, yellowish
brown spores of
Coniophora puteana

40 mm

8 μm

Adult fruit-body of
Fibuloporia vaillantii

Elongate ellipsoid,
hyaline spores of
Fibuloporia vaillantii

Shallow pores

25 mm

20 μm

Hyphae of *Fibuloporia vaillantii*

(c) Colonisers of cones

Auriscalpium vulgare S. F. Gray   Ear-pick fungus
Cap: 8–12 mm. *Stem:* width 4–6 mm; length 40–75 mm.
Description:
Cap: kidney-shaped or semicircular, thin, date- or umber-brown, hairy, but paler towards the margin.
Stem: erect, slender, hairy, particularly at the base, and attached at the side of the cap (excentric).
Gills: replaced by flesh-coloured, then greyish brown spines.
Flesh: brown.
Spore-print: white.
Spores: small, hyaline, minutely spiny, spherical, 4–5 μm in diameter, and becoming blue-grey in solutions containing iodine.
Cystidia: flask-shaped with oily contents.
Habitat & *Distribution:* This fungus is always found on fallen pine-cones and occurs from early summer to autumn.
General Information: The ear-pick fungus is easily recognised by the slender, elegant habit, excentrically placed cap, substrate preference and dark colours. It cannot be confused with any other fungus. Recently it has been shown that the 'agaric' *Lentinellus cochleatus* (Fries) Karsten (p. 76) is more closely related to *Auriscalpium* than this fungus is to other spine-bearing forms and *Lentinellus* is to the other agarics. Both fungi possess thick-walled cells in the flesh and oil-containing hyphae; they are placed in the family *Auriscalpiaceae*.

Another laterally stemmed Hedgehog fungus differs in possessing distinctly gelatinised teeth and preference for conifer wood and not cones. Examination of the basidia of this fungus shows that it is more closely related to the jelly-fungi, *Exidia* and *Tremella* (p. 184) than to Hedgehog fungi such as *Auriscalpium* or *Hyndum repandum* Fries (p. 160). This false nature is reflected in the name of the genus to which it belongs, *Pseudohydnum*, and the very gelatinous texture in the specific name '*gelatinosum*': the fungus is *Pseudohydnum gelatinosum*, or as it used to be called *Tremellodon gelatinosum*.

Illustrations: Auriscalpium vulgare—WD 103[6]. *Pseudohydnum gelatinosum*—WD 105[9].

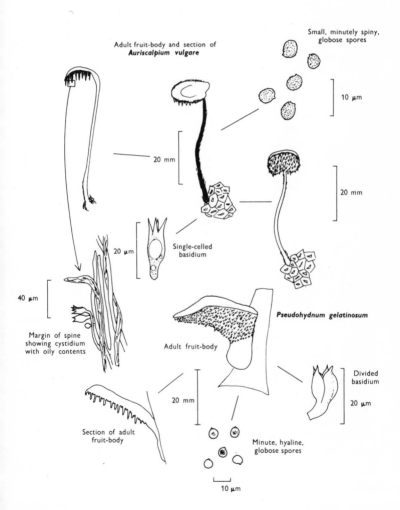

Small, minutely spiny, globose spores

Adult fruit-body and section of **Auriscalpium vulgare**

10 μm

20 mm

20 mm

Single-celled basidium

20 μm

40 μm

Margin of spine showing cystidium with oily contents

Pseudohydnum gelatinosum

Adult fruit-body

Divided basidium

20 μm

Section of adult fruit-body

20 mm

Minute, hyaline, globose spores

10 μm

(d) Terrestrial forms

Hydnum repandum Fries   Wood-hedgehog
Cap: 50–75 mm width. *Stem:* width 10–17 mm; length 45–65 mm.
Description:
Cap: rather thick, fleshy, pinkish buff or tan, paler at its incurved
   and often lobed margin.
Stem: short, stout and powdered with white roughenings and often
   attached to the cap to one side of the centre.
Gills: replaced by awl-shaped, pinkish buff spines which are unequal
   in length and run down the top of the stem.
Flesh: white, firm and with a pleasant smell.
Spore-print: whitish.
Spores: medium sized, hyaline under the microscope, smooth, broadly
   ellipsoid, 7×6–7 μm, and not becoming bluish grey in solutions
   containing iodine.
Cystidia: absent.
Habitat & Distribution: The 'wood-hedgehog' grows on the ground
   in mixed woods and is easily recognised by its colour and fleshy
   texture.
General Information: The closely related, smaller, red-brown species
   H. rufescens Persoon grows with conifers. *Hydnum* was formerly a
   genus which contained several entities, now not considered closely
   related. Thus the following genera have been delimited in addition
   to those related to *Hydnum repandum* and *H. rufescens*, and *Auriscalpium* described on p. 158.
       Sarcodon: Fruit-body fleshy: spores brown and ornamented
   with irregular bumps, e.g. *S. imbricatum* (Fries) Karsten.
       Phellodon: Fruit-body tough and fibrous: spores white and
   ornamented with small spines, e.g. *P. niger* (Fries) Karsten.
       Hydnellum: Fruit-body tough and fibrous: spores brown and
   ornamented with irregular bumps and bosses, e.g. *H. scrobiculatum*
   (Secretan) Karsten.
       Bankera: Fruit-body fleshy: spores white and ornamented with
   small spines, e.g. *B. fuliginoalbum* (Fries) Pouzar.
Illustrations: Hvass 280; LH 61; NB 153[3]; WD 53[4]; Z 61.

Plate 53. Tough or leathery fungi: Spores whitish and borne on spines

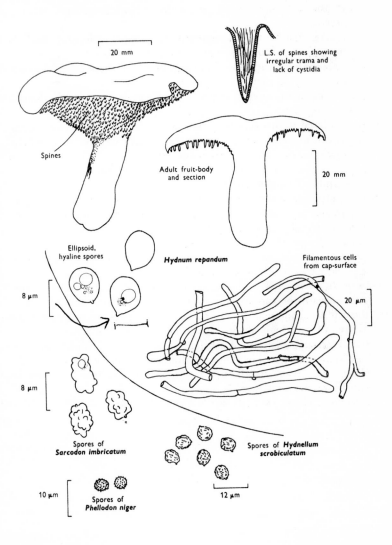

20 mm

L.S. of spines showing
irregular trama and
lack of cystidia

Spines

Adult fruit-body
and section

20 mm

Ellipsoid,
hyaline spores

Hydnum repandum

Filamentous cells
from cap-surface

8 μm

20 μm

8 μm

Spores of
Sarcodon imbricatum

Spores of *Hydnellum
scrobiculatum*

10 μm

Spores of
Phellodon niger

12 μm

(ii) Chanterelles and relatives

Cantharellus cibarius Fries   Chanterelle
Cap: 30–100 mm. *Stem:* width 15–25 mm; length 30–70 mm.
Description:

Cap: convex then flattened, irregularly wavy, more or less top-shaped,
   depressed and smooth or slightly roughened at centre, egg-yellow
   or lemon-chrome with flush of orange and with the margin incurved
   at first.

Stem: short, stout, tapered downwards, fleshy and similarly coloured
   to the cap.

Gills: replaced by irregularly branched yellow folds which may form
   a network near the margin and at the apex of the stem over which
   the folds run down irregularly (decurrent).

Flesh: with pleasant, fruity smell, yellow but paler on drying.

Spore-print: pale cream-colour.

Spores: medium sized, ellipsoid, thin-walled, smooth, 8–10 × 5–6 μm
   in size and not becoming bluish grey in solutions containing iodine.

Marginal and facial cystidia: absent.

Basidia: 2–8 spored.

Habitat & Distribution: Very common in troops in deciduous woods
   especially those with beech and oak.

General Information: Easily recognised by its folds and absence of true
   gills beneath the cap and the overall yellow colour. This fungus is
   the edible chanterelle of the continental market, where it is considered of very high quality; it can be purchased in this country in
   tins. *C. friesii* Quélet is of a bright apricot colour with lilaceous or
   rose-coloured flesh. The 'false chanterelle' *Hygrophoropsis aurantiaca*
   (Fries) Maire already discussed (see p. 106) has true gills and is
   reddish orange in colour.

Illustrations: Hvass 182; LH 59; NB 123[2]; WD 83[1].

Plate 54. Fleshy but firm fungi: Spores pale-coloured and borne on irregular folds (False gills)

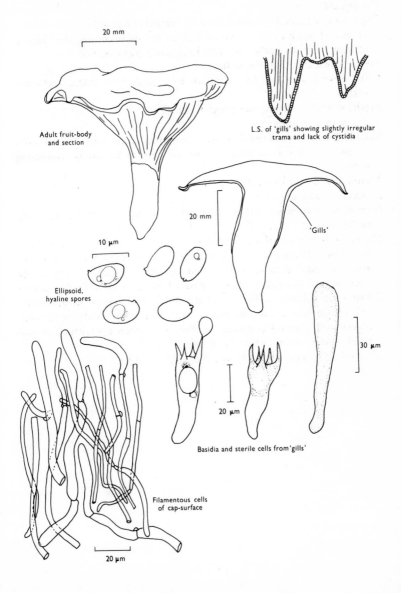

20 mm

Adult fruit-body
and section

L.S. of 'gills' showing slightly irregular
trama and lack of cystidia

20 mm

'Gills'

10 μm

Ellipsoid,
hyaline spores

30 μm

20 μm

Basidia and sterile cells from 'gills'

Filamentous cells
of cap-surface

20 μm

Craterellus cornucopioides (Fries) Persoon   Horn of plenty
Cap: 22–80 mm. *Stem:* width 15–25 mm; length 25–80 mm.
Description:
Cap: funnel-shaped, membranous to leathery, but limp, dark brown
   or almost black in wet weather, but on drying becoming dull brown
   or sepia, slightly scaly and with irregularly wavy margin.
Stem: short, blackish and hollow.
Gills: absent, replaced by a smooth to irregularly wrinkled, ash-grey
   surface.
Flesh: sepia but drying out greyish ochre.
Spore-print: cream-colour.
Spores: medium sized, hyaline under the microscope, ellipsoid, smooth,
   10–11 × 6–7 μm in size and not blueing in solutions containing
   iodine.
Marginal and facial cystidia: absent.
Basidia: usually 2-spored.
Habitat & *Distribution:* Often in very large troops in woods, especially
   under beech.
General Information: This fungus is recognised by the peculiar shape
   and dull colours which conceal it so well amongst the dead leaves
   and woodland debris; in the shade of the tree-canopy it is easily
   overlooked. *Craterellus sinuosus* (Fries) Fries is a much smaller species
   with dirty ochraceous fertile surface and brownish grey cap and stem.
   'Cornucopioides' means like (oides) a horn of plenty, a familiar
   object in mediaeval paintings as part of heathen festivities full and
   overflowing either with fruit or wine, or both!
Illustrations: Hvass 186; LH 59; NB 123[1]; WD 83[4].

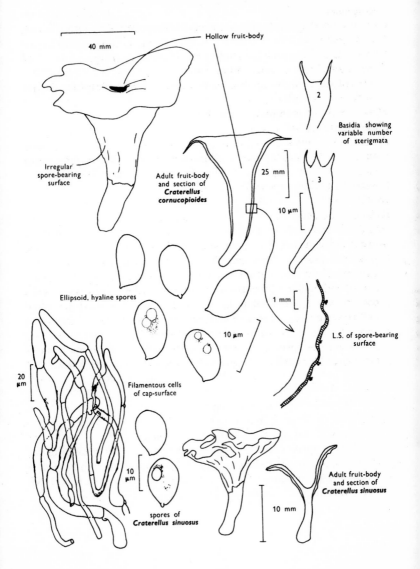

Hollow fruit-body

40 mm

2

Basidia showing variable number of sterigmata

Irregular spore-bearing surface

Adult fruit-body and section of *Craterellus cornucopioides*

25 mm

3

10 μm

Ellipsoid, hyaline spores

1 mm

L.S. of spore-bearing surface

10 μm

20 μm

Filamentous cells of cap-surface

10 μm

10 mm

10 μm

spores of *Craterellus sinuosus*

Adult fruit-body and section of *Craterellus sinuosus*

(iii) Fairy-club fungi

Clavulina rugosa (Fries) Schroeter   Wrinkled club
Cap: absent. *Fruit-body:* length 50–100 mm; width 7–13 mm.
Description:
Fruit-body: club-shaped, simple with blunt apex or irregular blunt
   branches, white or dirty cream colour, often thickened upwards and
   marked with longitudinal wrinkles or grooves and the whole surface
   of the club bearing spores.
Stem: absent or extremely short.
Flesh: white.
Spore-print: white.
Spore: medium sized, broadly ellipsoid to subglobose, hyaline under
   the microscope and not turning bluish grey in iodine solutions,
   9–10 × 7–8 μm in size.
Cystidia: absent.
Basidia: 2-spored.
Habitat & Distribution: Frequent on the ground in woods, especially
   in the shade of beech trees or in conifer plantations.
General Information: Two very closely related species are to be found
   in similar localities and are equally as common; they are *C. cristata*
   (Fries) Schroeter with strongly branched white fruit-body, each
   branch ending in pinkish or lavender-white, divided, sharply pointed
   branchlets and *C. cinerea* (Fries) Schroeter with irregular greyish or
   dark grey branches with a flush of violaceous.
      These three species are very closely related; in fact so many inter-
   mediates between the extreme morphological forms are known that
   some authorities have considered them simply forms of a single
   species. All these species lack cystidia.
         rugosa—wrinkled, referring to the spore-bearing surface.
         cristata—crested, referring to the branchlets.
         cinerea—ash-grey, referring to the colour.
      All these species are often found blackened by the growth of the
   microscopic fungus, *Helminthosphaeria clavariae* (Tulasne) Fuckel.
Illustrations: C. *rugosa*—LH 55; WD 104[5]. C. *cristata*—LH 55; NB
   153[5]; WD 104[2]. C. *cinerea*—WD 104[1].

All fruit-bodies same size

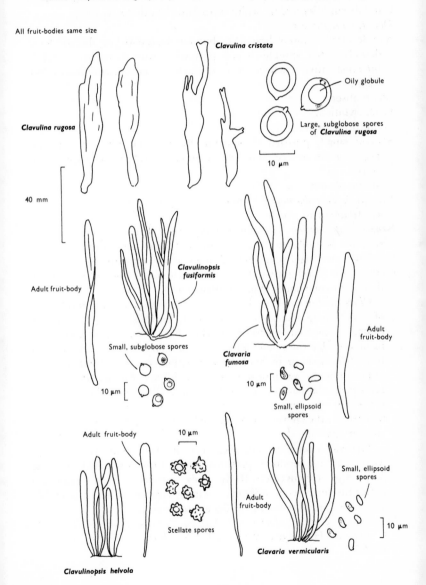

Clavulina cristata

Clavulina rugosa

Oily globule

Large, subglobose spores
of *Clavulina rugosa*

10 μm

40 mm

Adult fruit-body

Clavulinopsis fusiformis

Small, subglobose spores

10 μm

Clavaria fumosa

Small, ellipsoid
spores

10 μm

Adult
fruit-body

Adult fruit-body

10 μm

Stellate spores

Adult
fruit-body

Small, ellipsoid
spores

10 μm

Clavaria vermicularis

Clavulinopsis helvola

Clavaria vermicularis Fries   White spindles
Cap: absent. *Fruit-body:* width 6–10 mm; length 50–85 mm.
Description: Plate 56.

Simple or very rarely branched, but not forked below the soil-level,
  densely tufted, spindle-shaped, pure white with sharp, often slightly
  brownish, tips, when old it is wavy, often twisted, compressed and
  fragile.

Stem: absent.

Flesh: whitish.

Spore-print: white.

Spores: small, pip-shaped, smooth, hyaline under the microscope,
  4–5 × 3 μm in size, and not becoming bluish grey in iodine solutions.

Cystidia: absent.

Habitat & Distribution: Common in autumn amongst grass in fields,
  less frequent in woods.

General Information: Clavulinopsis fusiformis (Fries) Corner, 'Golden
  spindles' is similar to *C. vermicularis*, but forms dense tufts of canary-
  yellow, very fragile clubs joined in 2's or 3's below the soil level; the
  spores are also slightly different, being almost globose, hyaline under
  the microscope and 5–7 μm in diameter.

Clavaria fumosa Fries is similar to *C. vermicularis* and forms tufts
  of very fragile mouse-grey clubs with brownish tips; it produces
  elongate ellipsoid spores measuring 6–8 × 3–4 μm which are
  hyaline under the microscope. *C. vermicularis* and *C. fumosa* differ
  from *Clavulinopsis* in hyphal construction, but the differences are
  rather difficult to demonstrate to the beginner. *Clavulinopsis helvola*
  favours similar habits to *C. fusiformis* and although yellow in colour
  differs in the more orange-yellow colouration, but more particularly in the spores being rounded, 5–6 μm in diameter with large
  angular spines.

The earth-tongues, i.e. members of the family *Geoglossaceae*
  which are also found in pastures belong to an unrelated group of
  fungi, the Ascomycetes. If the clubs are crushed and examined
  under the microscope rows of sacs (asci) containing long thread-like
  ascospores are found—no basidia are to be seen.

Illustrations: Clav. fusiformis—WD 104[9]. *C. vermicularis*—WD 104[10].
  C. fumosa—Hvass 303; WD 104[11]. *Clav. helvola*—Hvass 300; WD
  105[1].

Plate 57. Club-shaped and coral fungi

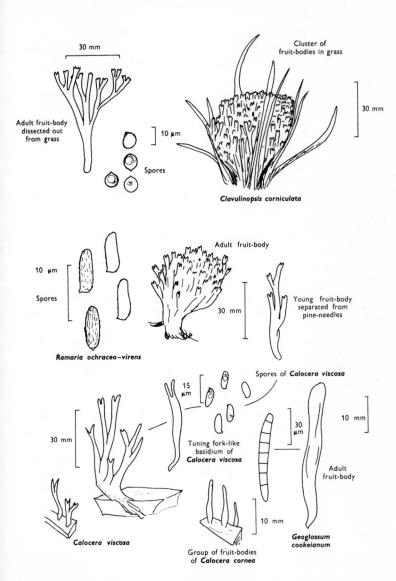

30 mm

Adult fruit-body
dissected out
from grass

10 μm

Spores

Cluster of
fruit-bodies in grass

30 mm

Clavulinopsis corniculata

10 μm

Spores

Ramaria ochraceo–virens

Adult fruit-body

30 mm

Young fruit-body
separated from
pine-needles

Spores of *Calocera viscosa*

15
μm

Tuning fork-like
basidium of
Calocera viscosa

30 mm

Calocera viscosa

30
μm

10 mm

Adult
fruit-body

*Geoglossum
cookeianum*

10 mm

Group of fruit-bodies
of *Calocera cornea*

Clavulinopsis corniculata (Fries).Corner (p. 171).

Cap: absent. *Fruit-body:* complex; width 20–30 mm; length 20–40 mm.

Description: Plate 57.

Fruit-body: shape depending on the length of grass in which it grows but always branching strongly from its base, composed of a dense compact tuft of egg-yellow or orange-tawny branches which are either irregular or of equal length and so they form a flattened top to the fruit-body complex, the branchlets are slender, forked 2- or 3-times, with their apices narrowed or curved.

Stem: very downy at the base.

Flesh: pale yellow.

Spore-print: white.

Spores: medium sized, hyaline under the microscope, smooth, spherical and 5–7 μm in diameter, not becoming bluish grey in iodine solutions.

Cystidia: absent.

Habitat & Distribution: Common amongst grass in fields or on grassy path sides in woodland.

General Information: Clavulinopsis corniculata is recognised by the branched habit and the smooth spores; *Ramaria ochraceo-virens* is of similar form, but has an overall duller colour and turns green on bruising, grows in pinewoods and has finely roughened brownish spores. *Calocera viscosa* also has an erect, bright golden or orange-yellow fruit-body which becomes more orange on drying. It is repeatedly branched and usually has a long, tough-rooting base. However, the spore-print is dirty yellowish and the fruit-body, which grows on coniferous wood, is viscid and elastic, a character reflected in the name 'viscosa'. Microscopically the basidium of *Calocera* is shaped like a tuning-fork and is not clavate as in *Clavulinopsis corniculata*. It appears to be more related to the jelly-fungi (see p. 180).

Illustrations: Clavulinopsis corniculata—LH 55; NB 6; WD 104[3]. *Calocera viscosa*—Hvass 304; LH 225; NB 149[3]; WD 107[8].

Typhula erythropus Fries.
Cap: absent. *Fruit-body* up to 20 mm high.
Description:
Fruit-body: upper fertile portion club-shaped and not more than half the length, white, surmounting a reddish brown, thread-like, often wavy or twisted stem which is attached at its base to an ellipsoid bead-like structure, called a sclerotium.
Spore-print: white.
Spores: oblong, smooth, hyaline under the microscope, 6–7 $\times$ 2 μm in size and not becoming bluish grey in iodine solutions.
Cystidia: absent.
Habitat & *Distribution:* Not uncommon on dead leaves and twigs or dead herbaceous stems.

Pistillaria micans Fries.
Cap: absent. *Fruit-body:* up to 10 mm high.
Description:
Club-shaped or oblong, rose-pink hardly differentiated from the similarly coloured stem, and arising at most from a small pad of filaments.
Spore-print: white.
Spores: broadly ellipsoid to pip-shaped, smooth, hyaline under the microscope, about 10 $\times$ 6 μm (8–11 $\times$ 5–7 μm) in size and not becoming bluish grey in iodine solutions.
Cystidia: absent.
Habitat & *Distribution:* Not uncommon on dead herbaceous stems and leaves, especially those in damp places.
Illustrations: T. erythropus WD 105 [10]. *P. micans* WD 105[7].

General notes on the club-fungi

Early mycologists believed that the club-shaped nature of the fruit-body was important in the classification of these fungi. Thus the Earth Tongues (*Geoglossum*, see Plate 57), the Stag's horn fungi and relatives (*Xylosphaera* see p. 204) both Ascomycete groups, the Dacry-mycetales (a group of jelly-fungi, see p. 180) and the true fairy-clubs were all classified together. It was 'the Father of Mycology', the Swede, Elias Fries, who in 1821, as in many other groups of fungi, made an attempt to make some sense of the chaos. By very careful observations, and what is so amazing without using a microscope, he was able to separate the tough stemmed and gelatinous stemmed groups from the more slender or coral-like ones. Fries was a very keen observer and noticed features which many modern authorities miss in the field because they rely too heavily on the microscope. Fries' system was used almost unchanged until the second half of this century; its beauty was its simplicity in that it joined together in one group all those fungi with simple basidia and the spore-bearing tissue distributed all around a simple club or around the branches of a complex fruit-body resembling a coral. However, by a careful examination of the micro-scopic structures, such as the spores and hyphae and the development of the fruit-body, it has been found necessary to separate these fungi still further. The ecology of the club-fungi has assisted in an under-standing of these proposed divisions.

The larger many branched clavarias, more correctly placed in the genus *Ramaria*, are to be found on bare soil in woodlands and planta-tions; *R. ochraceo-virens* is common in conifer plantations and can be recognised by the long ornamented spores, which characterise this group of fungi, and the fact that the sandy-coloured fruit-body be-comes dark olive-green on bruising (see p. 170).

Clavariadelphus pistillaris is the largest of our simple club-fungi; it may grow up to 200 mm high and 50 mm wide. This fungus has a wrinkled outer surface and sometimes the apex of the club becomes flattened and lacks basidia; this suggests a possible relationship, perhaps evolutionary, to the primitive chanterelles (see p. 162)—also woodland fungi. *Clavulina*, a complex group of dull or whitish, branched fruit-bodies, has been described earlier and the genus is characterised by the large spores and 2-spored basidia; they are woodland fungi also.

The grassland species are often simple in structure belonging in the main to the genus *Clavulinopsis* (see p. 170) and the now much reduced genus *Clavaria* (see p. 168). Although really complex, some of these species of *Clavulinopsis* are branched only below the soil level and thus appear as single clubs amongst the grass. Perhaps the single club has evolved especially to grow amongst blades of grass. *C. corniculata*, however, is well branched and the head is tight and compact and often flattened close to the ground. The same fungus in woodland is more open and because of this it was thought to be a different species to the grassland form. It is the simple club which dominates the form of those species which grow on herbaceous debris and grass-stems; indeed several species of *Typhula* cause diseases of grass particularly those of lawns where they have suffered damage because of cold or long periods under the snow. Some of these small fungi produce a small hard mass of fungal tissue about the size of a lupin seed (called a sclerotium). This is a resting body from which the club-shaped almost filament-like fruit-body later develops.

Thelephora terrestris Fries　　　　　　　　　　　　Earth-fan
Cap: absent. *Fruit-body:* width 20–40 mm; height 30–50 mm.
Description:
Fruit-body: erect, fan-shaped or effused with upturned margin, tough
　　but thin and fibrous, chocolate-brown or cocoa-coloured, scaly from
　　radiating fibrils and with fringed, pale buff or wine-coloured margin.
Gills: absent and replaced by a wrinkled or irregularly granular, dark
　　lilaceous grey or cocoa-coloured surface.
Flesh: brown and thin.
Spore-print: purplish brown.
Spores: medium sized, dark brown under the microscope, warted-
　　angular and 8–9 × 6–7 μm in size.
Cystidia: absent but basidia often filled with brown contents.
Basidia: 2–4 spored.
Habitat & *Distribution:* Found on the ground in woods, especially
　　pine woods; also on heathland growing up vegetation and incor-
　　porating it into the fruit-body's shape.
General Information: There is some evidence to suggest that this
　　fungus can form mycorrhiza with pine trees under certain con-
　　ditions.

　　Although it may be easily passed over because it is perfectly
camouflaged it is quite easy to recognise when collected. *T. palmata*
(Bulliard) Patouillard, is a bigger, less frequently seen species more
coral-like in shape; it also grows in pine woods. When the fruit-
body of *T. terrestris* spreads over the soil or plant debris it resembles
other members of the family to which it belongs, i.e. *Thelephoraceae*;
species of *Tomentella*. They also have warty angular spores, purplish
brown colours, and wrinkled or puckered spore-bearing surfaces.
Tomentella spp., however, are resupinate or encrusting and so do
not form caps, even at the margin of the fruit-body. *Tomentella* is
one of the many genera which were classed collectively as resupinate
fungi because they lack a cap and form crusts. This group 'the
resupinates' consists of a whole series of quite unrelated fungi.
Illustrations: LH 53; NB 47[8].

Plate 58. Club and Fan-shaped fungi

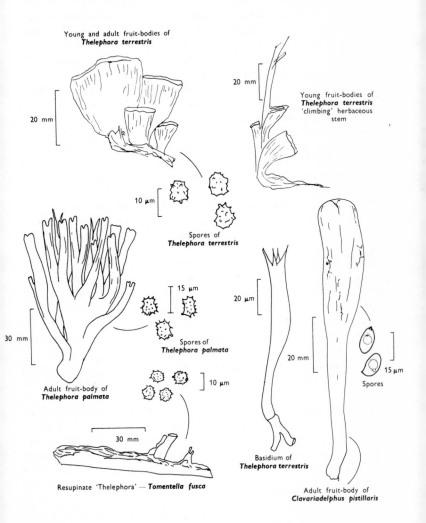

Young and adult fruit-bodies of
Thelephora terrestris

20 mm

Young fruit-bodies of
Thelephora terrestris
'climbing' herbaceous
stem

20 mm

10 μm

Spores of
Thelephora terrestris

15 μm

30 mm

Spores of
Thelephora palmata

20 μm

Adult fruit-body of
Thelephora palmata

10 μm

20 mm

15 μm

Spores

Resupinate 'Thelephora' — *Tomentella fusca*

30 mm

Basidium of
Thelephora terrestris

Adult fruit-body of
Clavariadelphus pistillaris

(iv) Resupinate fungi

When mycologists talk generally about 'resupinates' they are referring to a whole group of Basidiomycetes whose spore-bearing layer is exposed, the cap highly reduced or completely lacking, and the fungus adhering to the surface of the substrate which may be soil, wood, grasses, etc., at the point which would have been the cap of an agaric. Probably members of the group are the most commonly seen yet it is one of the most commonly ignored groups of fungi—by naturalists and mycologists alike; they form 'white wash' on old sticks, dark coloured discolourations on trunks, etc. It is an entirely artificial group of many quite unrelated elements united on the common factor of having either a reduced or primitive fruit-body consisting only of a sheet of tissue. However, these same fungi have a uniting factor in that they frequent the same ecological sites, e.g. on muddy soil in bogs, under overhangs of banks and stream sides, undersides of logs, trunks, branches and twigs, hidden in cracks of old stumps or spreading over carpets of conifer needles or dead leaves and sedges.

By studying the anatomy of the fruit-body and the characters of the spores certain relationships can be found which relate many of these fungi to several other groups of fungi we have dealt with in earlier chapters.

It is only possible to mention here the group as a whole for all the species really require very careful examination, often necessitating several hours of microscope work. They should be left by the beginner until more experience is obtained and advice of an expert easily available.

Although the group mainly contains saprophytes, a few are parasitic. 'Silver-leaf disease' of almonds and fruit trees is caused by *Stereum* (*chondrostereum*) *purpureum* (Persoon) Fries; it has a purple fruiting surface, and greyish upper surface when ever this is formed at the margin.

There are several species of *Stereum* in Britain, three species of which when handled in the fresh state stain red: *S. sanguinolentum* (A. & S.) Fries, a pale coloured species on conifer wood, *S. rugosum* (Pers.) Fries a similar coloured species on beech, birch and especially hazel, and *S. gausapatum* Fries an ochraceous yellow species on oak, often forming a pocket rot of the timber. However, the commonest member of the genus is an orange-tawny coloured species with a greyish buff,

Plate 59. Resupinate fungi

Microscopic structures at various magnifications

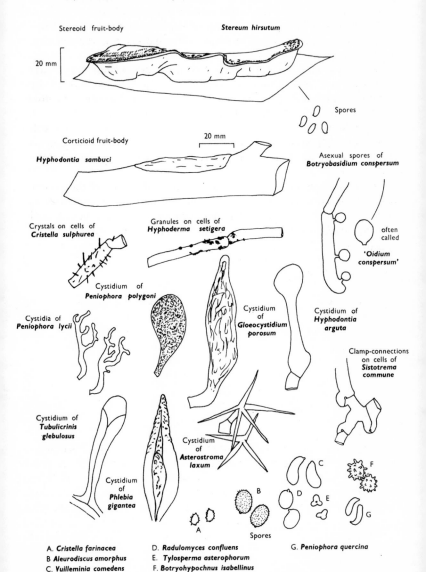

Stereoid fruit-body · **Stereum hirsutum**

20 mm

Spores

Corticioid fruit-body

Hyphodontia sambuci

20 mm

Asexual spores of
Botryobasidium conspersum

often
called

'**Oidium
conspersum**'

Crystals on cells of
Cristella sulphurea

Granules on cells of
Hyphoderma setigera

Cystidium of
Peniophora polygoni

Cystidia of
Peniophora lycii

Cystidium
of
**Gloeocystidium
porosum**

Cystidium of
**Hyphodontia
arguta**

Clamp-connections
on cells of
**Sistotrema
commune**

Cystidium of
**Tubulicrinis
glebulosus**

Cystidium
of
**Asterostroma
laxum**

Cystidium
of
**Phlebia
gigantea**

A

B

C

D

E

F

G

Spores

A. **Cristella farinacea**
B. **Aleurodiscus amorphus**
C. **Vuilleminia comedens**

D. **Radulomyces confluens**
E. **Tylosperma asterophorum**
F. **Botryohypochnus isabellinus**

G. **Peniophora quercina**

hairy cap, *S. hirsutum* (Willd) Fries. It grows on many trees of broad-leaved wood and can be found wherever twigs, branches, trunks or stumps have been lying out in the rain; it does not bleed.

Those species of resupinate fungi which resemble members of this genus, i.e. those with a distinct tough, although poorly developed, cap, are called stereoid.

'Red thread disease' of grass which often causes unsightly red patches on lawns and school and corporation playing-fields is caused by *Corticium fuciforme* (Berkeley) Wakefield. Fungi belonging to this genus produce fruit-bodies which 'scramble' over the substrate; for example, if one searches old elder trees throughout the year one will certainly find a 'white wash' fungus of this type, *Hyphodontia sambuci* (Pers.) J. Eriksson. Fungi with this type of fruit-body are called corticoid.

The two major types are illustrated along with some of the bizarre microscopic structures one finds in the resupinates; such structures are useful in classification and identification, and their beauty and intricacy make up for the surprisingly simple fruit-body shape and texture.

C. THE JELLY FUNGI

Key to the major groups

The jelly fungi or Hymenomycetous heterobasidiae is a complex group of fungi and not only includes our common jelly fungi but many microscopic forms some of which are parasitic. The group is divided into three main divisions depending on the position of the cross-walls within the basidium, or whether the basidium is in the shape of a tuning-fork. They are probably not closely related one to another.

Auriculariales (Basidia divided into cells by transverse walls)
1. Fruit-body lacking a cap and more or less forming a gelatinous coating on plant-debris .. .. *Helicobasidium*
   Fruit-body with more or less distinct cap, gelatinous but tough 2
2. Fruit-body ear-like or cup-shaped; upper surface with grey hairs and lower surface lilaceous brown or wine-coloured *Hirneola*
   Fruit-body at first cup-shaped but then spreading; upper surface grey and hairy, and lower surface purplish. *Auricularia*

Tremellales (Basidia divided into cells by longitudinal walls)
1. Fruit-body with distinct stem and spines on lower surface
   Pseudohydnum
   Fruit-body lacking a well-developed stem, either reduced to a small lobe or entirely absent 2
2. Fruit-body flattened or disc-shaped, often with warts or veins on the surface; spores more or less sausage-shaped. .. *Exidia*
   Fruit-body brain-like or with irregular bumps, sometimes lobed or irregular and encrusting 3
3. Fruit-body brain-like or with bumps or bosses; spores rounded or ovoid .. .. *Tremella*
   Fruit-body encrusting woody or herbaceous material; spores ellipsoid .. .. *Sebacina*

Dacrymycetales (Basidia resembling the shape of a tuning-fork)

1. Fruit-body club-shaped or coral-like   ..   ..   ..   *Calocera*
   Fruit-body top-shaped or with irregular bumps   2
2. Fruit-body top-shaped   ..   ..   *Femsjonia*
   Fruit-body cushion- or brain-like, or with irregular bumps
                                                        Dacrymyces

Dacrymyces stillatus Nees ex Fries

Fruit-body: 3–6 mm.

Description:

Fruit-body: cushion or brain-like, often irregular, lacking any evidence of stem, yellow or orange, gelatinous, covered entirely by spore-bearing tissue.

Spore-print: yellowish.

Spores: long, cylindrical or oblong, and slightly curved and $12–15 \times 5–6$ μm in size; they characteristically have 2 to 4 cross-walls dividing the interior of the spore (see below).

Cystidia: absent.

Habitat & *Distribution:* Common on all sorts of old wood, particularly on fence-posts, wooden railway-sleepers and other worked timber outside, e.g. sides of summer-houses and garden sheds. It is also found on twigs and branches in woods and copses.

General Information: This fungus is found throughout the year, but it is much more obvious under damp conditions when it is strongly gelatinised and very soft; when dry it almost disappears. The tissue bearing the basidia (perfect state) is yellow, when orange there is a predominance of asexually produced spores called arthrospores (conidia).

D. *deliquescens* is only another name for the same fungus. There are several species of *Dacrymyces* with which *D. stillatus* can be confused, but can only be separated with certainty by using a microscope. The Coral-spot fungus, frequently found in gardens, produces gelatinous, pink protuberances on wood especially that of sycamore, and may easily be mistaken for species of *Dacrymyces*. It consists entirely of asexually produced spores (conidia) of the Ascomycete *Nectria cinnabarina*. The perfect state appears late in the year as grouped, small, blood-red flask-shaped fruit-bodies containing envelopes of spores. It is quite unrelated to *Dacrymyces*.

THE JELLY FUNGI 181

Calocera viscosa (Fries) Fries described earlier (p. 170) is closely related to *Dacrymyces*. The much smaller, and probably equally as common, *Calocera cornea* (Fries) Fries is simple, club-shaped and yellow, but darkens to become orange on drying. It grows up to 10 mm high and occurs on all sorts of wood; it is especially common on wet beech trunks. It approaches *Dacrymyces* more than the much larger *C. viscosa*.

Illustrations: D. deliquescens—LH 225; NB 149[7]; WD 107[10]. *C. cornea*—WD 107[9].

Hirneola auricula-judae (St Amans) Berkeley   Jew's ear
Fruit-body: width 20–75 mm.
Description:
Fruit-body: cup or ear-shaped, red-brown or deep wine-colour, gelatinous with its upper surface, velvety and clothed in greyish or olivaceous hairs.

Spore-bearing layer: reddish or purplish brown, smooth or veined and translucent.

Spore-print: white.

Spores: very long, hyaline under the microscope, oblong, curved and narrowed towards their base, 16–18 × 6–8 μm in size.

Cystidia: absent.

Habitat & *Distribution:* On dead branches of all kinds and particularly common throughout the year on elder.

General Information: Easily recognised by the wine-coloured, cup-shaped or ear-shaped fruit-body; it is often called *Auricularia judae* in many books. Its Latin name is reflected in the common name:— *auricula* ear and *judae*, of a jew. This fungus is supposed to be a reappearance, as a warning to us all, of Judas, who on betrayal of Christ hung himself from an elder tree.

Auricularia mesenterica (S. F. Gray) Persoon, 'Tripe-fungus', is bracket-shaped with a hairy upper surface and reddish purple or deep purple lower surface which when fresh has a greyish bloom due to the formation of the spores.

    There are several fungi in the group Auriculariales in Britain, but many of them are inconspicuous and are identified with difficulty except by the expert. *Sebacina incrustans* (Fries) Tulasne is a common more obvious example of the resupinate forms. It grows as a cream or ivory-coloured, soft fruit-body encrusting twigs, leaves, grass and soil.

Illustrations: LH 225; NB 149[1]; WD 107[1].

Plate 60. Jelly fungi

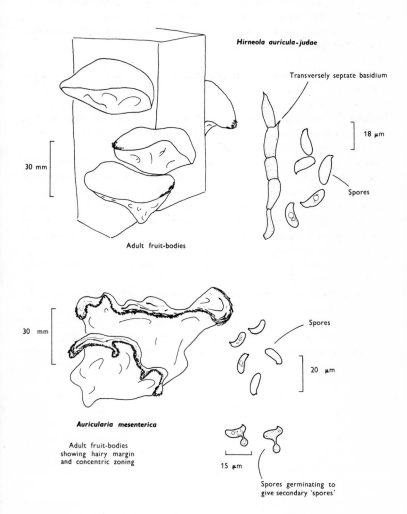

Hirneola auricula-judae

Transversely septate basidium

18 μm

Spores

30 mm

Adult fruit-bodies

30 mm

Spores

20 μm

Auricularia mesenterica

Adult fruit-bodies
showing hairy margin
and concentric zoning

15 μm

Spores germinating to
give secondary 'spores'

Exidia glandulosa (St Amans) Fries   Witch's butter
Fruit-body: width 15–50 mm.
Description:
Fruit-body: sessile or shortly stalked, blackish, variable in shape, rounded, flattened, disc-shaped or convolute, gelatinous with its under surface tomentose and free from the substrate.
Fruiting surface: upper most, wavy and folded, and with numerous wart-like projections.
Spore-print: white.
Spores: long, hyaline, cylindrical, sausage-shaped and 12–15 × 5 μm in size.
Cystidia: absent.
Habitat & *Distribution:* Frequent in crowded groups on stumps, logs and fallen branches of broad-leaved trees, especially those of ash; common throughout the year.
General Information: Tremella foliacea (S. F. Gray)Persoon and *Tremella mesenterica* Hooker are similar but more convoluted with leaf-like lobes. The former is cinnamon brown and occurs on conifer wood and its spores are 7–9 × 5–7 μm, whilst the latter is bright golden yellow or orange-yellow and occurs on broad-leaved trees. *T. mesenterica* has spores 7–8 × 5–6 μm, often accompanied or replaced by small, asexually produced spores.

Glandulosa—means full of glands and refers to the glands of the upper surface of the Witch's butter.

The convoluted fruit-body of the *Tremella* spp. is reflected in the word foliacea—leafy, and mesenterica—middle intestine. The last species is also often called the 'Yellow brain-fungus'.
Illustrations: Exidia glandulosa—LH 225; NB 149[4]; WD 107[3]. *Tremella mesenterica*—LH 225; NB 149[5]; WD 107[6].

Plate 61. Jelly fungi

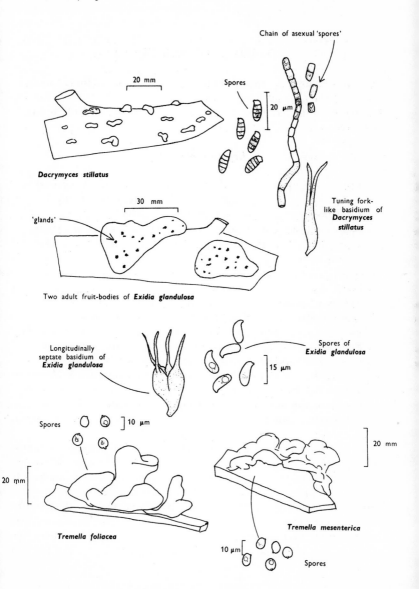

Chain of asexual 'spores'

Spores

20 mm

20 μm

Dacrymyces stillatus

Tuning fork-like basidium of *Dacrymyces stillatus*

'glands'

30 mm

Two adult fruit-bodies of *Exidia glandulosa*

Longitudinally septate basidium of *Exidia glandulosa*

Spores of *Exidia glandulosa*

15 μm

Spores

10 μm

20 mm

Tremella foliacea

20 mm

Tremella mesenterica

10 μm

Spores

D. THE STOMACH FUNGI

The Gasteromycetes are a complex mixture of higher fungi united in virtue of their spores being enclosed in a fruit-body and not forcibly ejected from the basidium; the group includes the puff-balls and their relatives.

Key to some groups

1. Fruit-body growing beneath the surface of the soil (hypogeous)
   False truffles (including *Hymenogaster*, *Rhizopogon*)
   Fruit-body not growing beneath the soil-surface   2
2. Spores in a slimy mass on a specialised fruit-body arising from
   an egg-like structure   ..   Stinkhorns (*Phallus* & *Mutinus*)
   Spores powdery at maturity or in small capsules   3
3. Spores powdery at maturity and contained within the fruit-
   body   4
   Spores enclosed in a small capsule or group of capsules in a
   cup-like structure, resembling the eggs within the nest of a
   bird ..   .. Bird's nest fungi (including *Crucibulum* & *Cyathus*)
4. Spores intermixed with threads within the fruit-body from
   which they are dispersed through a specialised pore at its apex
   Puff-balls and Earth-stars (*Lycoperdon* & *Geastrum*)
   Spores not mixed with threads within the fruit-body and not
   dispersed through special structure but through cracks as the
   fruit-body weathers   ..   ..   Earth-Balls (*Scleroderma*)

The Gasteromycetes is an unnatural group of predominantly sapro-phytic higher fungi many of which are extremely grotesque and strange in their morphology. Instead of the spores being formed asymmetrically on the basidium as is found in the agarics, the spores of members of this group are usually more or less symmetrically attached to their sterig-mata or may even be seated directly (sessile) on the basidium. The whole group, even if unnatural, can, however, be regarded under one heading as a biological unit. Until something better is suggested and supported by evidence the existence of this group is very convenient.

Usually the basidia project into cavities within the fruit-body in which the spores themselves are released as the fruit-body gradually matures—hence the name Gastero-mycetes: 'stomach-fungi'. In a few more advanced forms, the puff-balls of temperate countries, for instance, the spores escape from these cavities through a pore or pores in the outer wall of the fruit-body, and in the stinkhorns the spores are exposed as a sticky mass because the smell of the material in which they are held is attractive to flies. In forms which have subterranean (or hypogeous p. 243) fruit-bodies there is no special opening and here the spores are dispersed by insects and small mammals. In the bird's nest fungi the spores are enclosed in separate packets within a saucer or cup-like open structure.

Recently it has been shown by examination of the microscopic structure of the fruit-bodies and spores that certain genera of the Gasteromycetes are more closely related to the agarics than many of them are between themselves.

It is believed that some of the Gasteromycetes may have evolved from more familiar fungi by adaptation to arid or semi-arid conditions. Although this is not true for all the Gasteromycetes within this one group of fungi, a whole series of methods of overcoming the disadvantages connected with non-violent disposal of spores has evolved. These methods include both changes in structure and ecology; only a few have evolved a mycorrhizal relationship with higher plants.

Lycoperdon pyriforme Persoon   Stump puff-ball
Fruit-body: width 20–50 mm; height 40–75 mm.
Description:

Fruit-body: more or less pear-shaped, pale brownish often with a
    slight hump on the top, scurfy on the outside with tiny pointed
    granules which soon fall off or become rubbed off by abrasion,
    particularly after careless handling.

Stem: consisting of rather small cells and connected at the base by
    long, white, branched cords of mycelium which permeate the
    substrate.

Spore-mass: white at first then greenish yellow and finally olive-brown
    and formed around a sterile column.

Spores: small, olive, minutely warted but appearing smooth under the
    student microscope; 4 μm in diameter and intermixed with long,
    olive coloured, branched hyphal threads 4–5 μm broad and of
    irregular wall thickness.

Habitat & Distribution: This puff-ball grows in huge clusters on stumps
    and logs, or can be traced to buried pieces of wood; it occurs from
    summer to late autumn.

General Information: There are several species of *Lycoperdon* in this
    country, some quite small and several rather infrequent. *L. pyriforme*
    is the only one which grows on wood; 'pyriforme' means pear-
    shaped and is derived from the shape of the fruit-body.

L. perlatum Persoon is also a common puff-ball; it is pestle-shaped
    or top-shaped, whitish or tan with minutely roughened, globose
    spores measuring 4 μm in diameter. The fruit-body is covered in a
    mixture of large and small, fragile spines which leave a network
    when rubbed off. It grows in woods and on heaths.

L. foetidum Bonorden is similar to *L. perlatum*, but the spines are
    umber or vandyke-brown; it also grows both in woods and upland
    pastures, particularly the latter.

Illustrations: L. pyriforme—Hvass 316; LH 219; NB 155[3]; WD 109[3].
    L. perlatum—Hvass 315; LH 217; NB 155[2]; WD 110[2].

Plate 62. Puff-balls

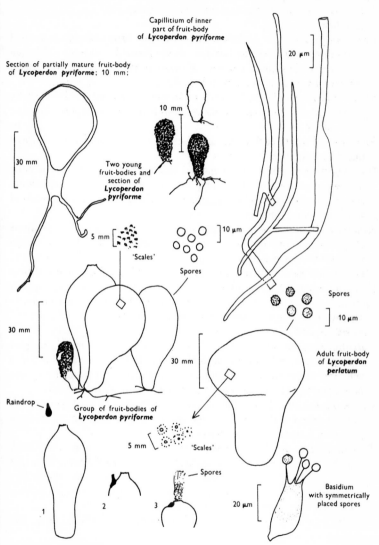

Capillitium of inner part of fruit-body of *Lycoperdon pyriforme*

20 μm

Section of partially mature fruit-body of *Lycoperdon pyriforme*; 10 mm;

30 mm

10 mm

Two young fruit-bodies and section of *Lycoperdon pyriforme*

5 mm   'Scales'

Spores   10 μm

30 mm

Spores

10 μm

Adult fruit-body of *Lycoperdon perlatum*

30 mm

Raindrop

Group of fruit-bodies of *Lycoperdon pyriforme*

5 mm   'Scales'

Spores

Basidium with symmetrically placed spores

20 μm

1   2   3

20 μm

1–3 Dispersal of spores
Falling raindrop depresses outer surface
and pushes out a cloud of spores

Langermannia gigantea (Persoon) Rostkovius   Giant puff-ball
Fruit-body: diameter 300–450 mm (–1,050 mm).
Description:
Fruit-body: round or slightly flattened on the top, smooth or cracked
    into small scales, white but becoming flushed yellowish with age
    and finally olive-brown when old, frequently the outer layer dries
    and breaks away to expose the powdery spore-mass within.
Stem: absent or only present as a small cone of tissue.
Spore-mass: whitish, cream-coloured and finally olive-brown.
Spores: small, brownish, minutely warted and spherical, 4–5 μm in
    diameter and intermixed with thick-walled, branched, brown
    hyphae, 3–5 μm broad.
Habitat & *Distribution:* On the ground in copses, at the edges of woods,
    under hedges or on refuse tips, and sometimes in gardens. It may
    appear in the same place year after year, and has been recorded
    growing beneath the rafters in houses.
General Information: When young it is white inside or cream-coloured
    before the spores have developed and can then be cut into slices
    and cooked. I have seen it on sale in markets in N. America and it
    is collected for food by many in Europe. Its pumpkin-shape with a
    circumference of anything up to 1,050 mm makes this fungus easily
    recognisable. The number of spores produced by a fruit-body
    measuring 400 × 280 mm has been calculated by A. H. R. Buller as
    7,000,000,000,000 spores!
Calvatia utriformis (Fries) Jaap (= *C. caelata* (Persoon) Morgan)
    has a goblet-like shape and a distinct, sterile base composed of large
    cells with a prominent membrane separating them from the spore-
    mass; the spores are 4–5 μm diameter, smooth and spherical.
C. excipuliformis (Fries) Perdeck (= *C. saccata* (Vahl.) Morgan) is
    pestle-shaped with a well developed stem. The spore-mass is com-
    posed of warted, globose spores, 4–5 μm in diameter.
Bovista nigrescens Persoon is very similar in shape to the Giant
    puff-ball, but is very much smaller; it lacks a stalk, being attached
    to the substrate only by mycelial cords. It commences white, but
    then darkens to become purplish brown at maturity when it also
    breaks from its moorings and rolls about in the wind.
    The last three species are found on heaths, in pastures or on the
ground in woods.
Illustrations: C. gigantea—Hvass 312; LH 217; NB 371; WD 109[7].
    B. nigrescens—Hvass 311; LH 219; NB 37[3].

Plate 63. Puff-Balls

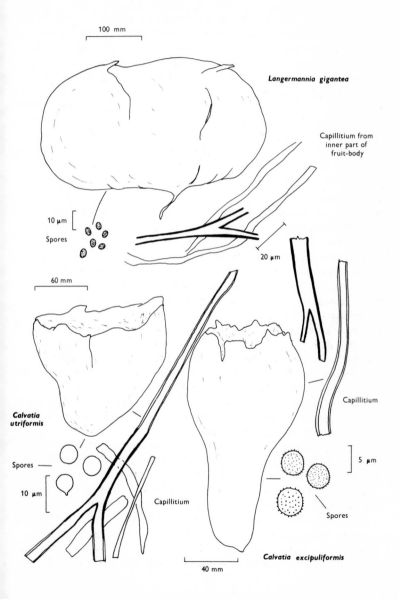

100 mm

Langermannia gigantea

Capillitium from
inner part of
fruit-body

10 μm
Spores

20 μm

60 mm

*Calvatia
utriformis*

Spores

10 μm

Capillitium

Capillitium

5 μm

Spores

40 mm

Calvatia excipuliformis

Earth-stars and Earth-balls

The earth-stars, i.e. species of *Geastrum*, are closely related to the puff-balls, but differ in having two very distinct and separate enclosing walls, the outer one splitting at maturity to expose a 'puff-ball' within; an example of the genus is *G. triplex* Jungh, found in parks or under beech trees or *G. rufescens* Pers. (illustrated) in mixed woodland. The outer skin splits in different ways in different species: in some it splits like a star—hence the common name of Earth-star, in some the spore-mass is raised as if on stilts. There are several species of *Geastrum* recorded for Britain, but they are decidedly uncommon.

The Earth-balls are, however, far from uncommon and may be met with from early summer until late autumn in any wood particularly those on sandy soils. They are unrelated to the earth-stars.

Earth-balls
Scleroderma citrinum Persoon   Common earth-ball
Fruit-body: diameter 25–75 mm.
Description:
Fruit-body: rounded or flattened on top, sometimes lobed, very firm, yellow or clay colour, scaly, thick, white within or pinkish, if cut when immature, and then purplish black as the spores mature.
Stem: absent or reduced to a small group of mycelial cords.
Spore-mass: purplish black.
Spores: medium to large, dark brown, 8–13 μm in diameter and covered with a delicate network.
Habitat & *Distribution:* On the ground in woods or on heaths.
General Information: This fungus is found in many books under the name of *S. aurantium*. *S. verrucosum* Persoon is closely related, but has a stem-like rooting base and an umber brown spore-mass. The spores are also slightly different; they are 10–14 μm in diameter and ornamented with spines and ridges.

The earth-balls appear to have characters in common with the false truffles, indeed sometimes they grow partially buried in the sandy soil of woods. Like the false truffles they have been used to adulterate pâté as a cheap substitute for true truffles (see p. 244). It is not wise, however, to eat earth-balls as there are cases of poisoning known. Although truffle-like, they should be avoided except under the guidance of an expert, as with agarics.
Illustrations: Geastrum triplex—Hvass 307; LH 221; NB 155[1].
Scleroderma citrinum—Hvass 320; LH 223; NB 155[5]; WD 111[3].

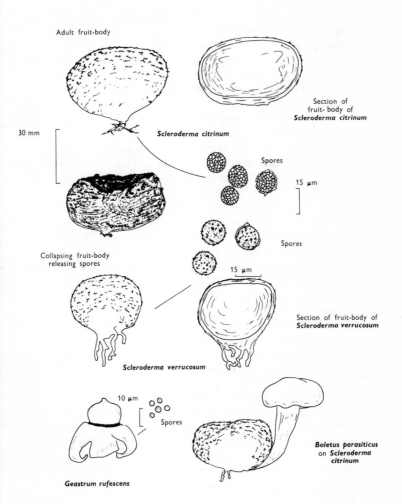

Plate 64. Earth-balls and Earth-stars

All fruit-bodies at same magnification

Adult fruit-body

Section of fruit-body of **Scleroderma citrinum**

Scleroderma citrinum

30 mm

Spores

15 μm

Collapsing fruit-body releasing spores

Spores

15 μm

Section of fruit-body of **Scleroderma verrucosum**

Scleroderma verrucosum

10 μm

Spores

Geastrum rufescens

Boletus parasiticus on **Scleroderma citrinum**

Stinkhorns

Phallus impudicus Persoon   Common stinkhorn

Fruit-body: Egg: 30–60 mm in diameter—then *Cap:* 25–40 mm and
  Stem: width 18–25 mm; length 100–150 mm.

Description:

Fruit-body: commencing as a white, silky egg-like structure full of
  jelly in which is embedded a conical cap attached only at its apex
  to a cylindrical white, spongy, hollow stem.

Cap: covered in a slimy mass of dark olive-coloured spores at maturity.

Stem: cylindrical, rapidly elongating, white, spongy and hollow.

Spore-mass: dark olive-green, smelling strongly, foetid.

Spores: small, pale olive, oblong and 3–5 $\times$ 2 μm in size.

Habitat & *Distribution:* Common from summer to autumn on the
  ground in woods and in gardens.

General Information: Easily recognised by its shape and evil smell
  which can be detected at some distance. The unburst eggs are
  called 'witches eggs'. Under favourable conditions the egg bursts
  and the stem elongates carrying the cap and spore-mass with it.
  The spore-mass is attractive to flies and they feed upon it; spores
  stick to their feet and so are transported from one place to another.

  The very similar *P. hadriani* Persoon is frequently found in sand-
  dunes; it differs in having a lilaceous colour to the egg. An interesting
  variety of the common stinkhorn is uncommonly found and differs
  in having a skirt-like frill beneath the cap. The jelly in the egg is
  a water-store and is used by the fungus to expand rapidly.

Mutinus caninus (Persoon) Fries, the 'Dog's stinkhorn', is found
  around old stumps or on piles of leaves. It has the spore-mass covering an orange-red pear-shaped cap which is itself fused to the stem.

  The stinkhorns and their allies appear to be commoner in warmer
  countries where they take on many bizarre shapes. Other than the
  three species noted above stinkhorns are rarely found in this country,
  but when they are it would appear they have been introduced with
  foreign imports such as timber, ornamental plants, vegetables etc.

  Eggs of phalloids brought into the laboratory can be surrounded
  by wet tissues or blotting paper and then allowed to develop further
  in a dish or box. Provided the skin covering the spores is not broken
  or injured the fungus will not smell and therefore before it becomes
  unpleasant, the whole mechanism of expansion can be studied.

Illustrations: Hvass 323; LH 215; NB 153[1]; WD 108[1].

Plate 65. Stinkhorns

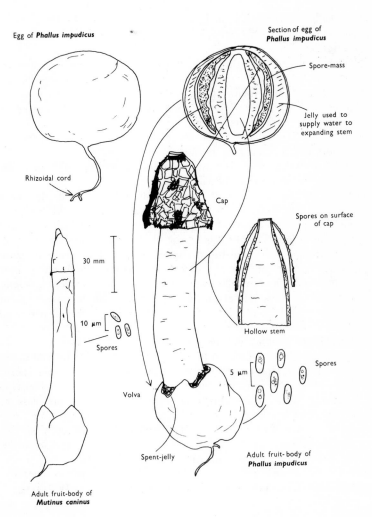

Egg of *Phallus impudicus*

Section of egg of *Phallus impudicus*

Spore-mass

Jelly used to supply water to expanding stem

Rhizoidal cord

Cap

Spores on surface of cap

30 mm

10 μm

Spores

Hollow stem

5 μm

Spores

Volva

Spent-jelly

Adult fruit-body of *Phallus impudicus*

Adult fruit-body of *Mutinus caninus*

Bird's nest fungi

Crucibulum laeve (de Candolle) Kambly
Fruit-body: diameter 8–12 mm.
Description:

Fruit-body: ochraceous brown or sand-colour, downy and then smooth, truncate, cup-shaped with the cup at first closed by a yellowish membrane which finally splits to expose a group of pale brown or dingy whitish, circular, lens-shaped 'eggs' (peridioles), scattered on a shiny pale ochraceous interior.

Spores: medium-sized, in packets within 'eggs', ellipsoid, hyaline, smooth and 8–10 × 4–6 μm in size.

Habitat & *Distribution:* Common in crowded groups on dead twigs, fern stems, straw and wheat stubble.

General Information: Cyathus differs from *Crucibulum* in the more complex fruit-body which consists of three layers, and the peridioles forming on distinct stalks. Two species are frequently seen: *Cyathus striatus* Persoon has a grey, fluted inner surface to the cup and strongly hairy red-brown outer surface; the spores measure 16–22 × 9–10 μm. *Cyathus olla* Persoon has a smooth, shiny, grey surface and minutely silky, yellowish grey outer surface. *C. striatus* is found on twigs, and about dead stumps; *C. olla* is more frequent in gardens on herbaceous debris and dead pieces of perennial flowers—or even in plant pots.

Sphaerobolus stellatus Persoon is more distantly related and grows on decaying leaves, bracken fronds, partially buried twigs and dung. It is an intriguing fungus because it possesses a remarkable spore-dispersal mechanism. The inner layer of the fruit-body when ripe suddenly turns inside out catapulting the inner spore-mass to distances of anything up to 4,200 mm, that is a distance of 1,000 times the size of the fruit-body. The fruit-body is externally whitish or pale yellow, but this layer splits into lobes like a star exposing the bright orange inner surface and pale spore-mass.

Illustrations: Crucibulum laeve—LH 223; WD 111[7]. *Cyathus striatus*—LH 223; WD 111[9]. *Sphaerobolus stellatus*—LH 223; WD 111[5].

Plate 66. Bird's nest fungi

Microscopic structures at various magnifications

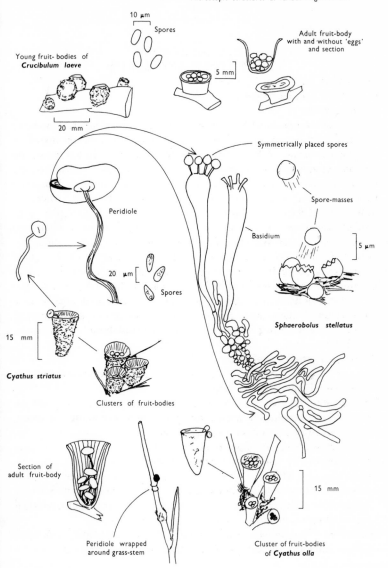

10 µm
Spores

Adult fruit-body
with and without 'eggs'
and section

Young fruit- bodies of
Crucibulum laeve

5 mm

20 mm

Symmetrically placed spores

Peridiole

Spore-masses

Basidium

5 µm

20 µm
Spores

Sphaerobolus stellatus

15 mm

Cyathus striatus

Clusters of fruit-bodies

Section of
adult fruit-body

Peridiole wrapped
around grass-stem

15 mm

Cluster of fruit-bodies
of **Cyathus olla**

E. CUP FUNGI AND ALLIES

General Notes.

The Ascomycetes differ from all the other fungi so far dealt with in that the spores develop enclosed in a microscopic envelope or sac—called the ascus. Usually eight spores are produced in each ascus and they are often dispersed violently into the air. Elf-cups and morels are typical Ascomycetes, but the group also includes innumerable minute forms of the microscopic fungi, small discs, minute flask-like structures, some of which are parasitic on leaves and stems of higher plants. In number the large species of Ascomycetes are few when compared with the others and therefore can only be given but a mention in the present account. When collected the Ascomycetes can be distinguished from the Basidiomycetes by simply examining a slice of the spore-producing tissue where the tell-tale asci will be seen (see p. 21). If the fruit-body is placed in a tin when collected and opened in a warm room all the mature asci explode at once producing a cloud of spores visible in the air immediately over the fruit-body.

Aleuria aurantia (Fries) Fuckel

Orange-peel fungus or Scarlet elf-cup

Fruit-body: diameter 25–50 mm.

Description:

Fruit-body: cup-shaped then undulating and becoming flattened, irregular, sometimes split and lacking a stem.

Inner surface: bright orange.

Outer surface: whitish and minutely downy.

Flesh: thin and white.

Spores: very long, ellipsoid, ornamented with a coarse network which projects at each end, and 17–24 × 9–11 μm in size; eight contained in an elongate, cylindric ascus.

Habitat & *Distribution:* Common on bare soil in woods and open spaces, on road verges, between stone sets and on lawns.

Plate 67. Cup-fungi

Adult fruit-body of *Aleuria aurantia*

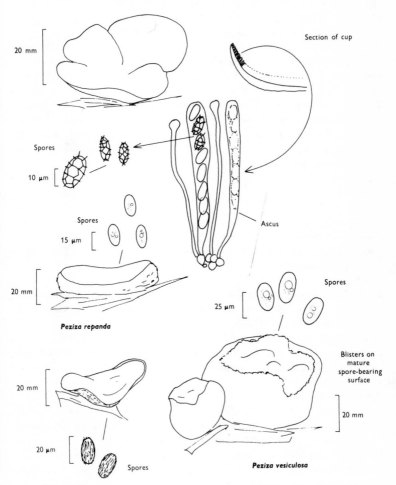

Section of cup

20 mm

Spores

10 μm

Spores

Ascus

15 μm

20 mm

Peziza repanda

Spores

25 μm

Blisters on
mature
spore-bearing
surface

20 mm

Peziza vesiculosa

20 mm

20 μm

Spores

Peziza badia

Peziza repanda Persoon   Elf-cup

Fruit-body: diameter 20–50 mm.

Description: Plate 67.

Fruit-body: cup-shaped with white, crenulate margin, becoming expanded and undulating, and lacking a stem.

Inner surface: light chestnut brown.

Outer surface: whitish or pale fawn and finely scurfy.

Flesh: whitish or fawn, and appearing as if layered.

Spores: very long, ellipsoid, smooth and 15–16 × 9–10 μm in size; eight contained in an elongate, cylindrical ascus.

Habitat & Distribution: On bare soil in woods, farm-yards, hedgerows, etc.

General Information: There are many different species of *Peziza* classified on the shape and ornamentation of the spores and colour of the fruit-body—see pp. 216 and 220. *P. badia* is darker, although similar in other ways; it is found on pathsides in woods and has roughened spores.

Morchella esculenta St Amans   Common morel

Cap: width 30–40 mm; length 35–60 mm. *Stem:* width 15–25 mm; length 50–80 mm.

Description:

Fruit-body: consisting of a head with a honeycomb-like arrangement of narrow ridges surrounding angular and often slightly elongated, shallow pits, on a cylindric or swollen stem.

Cap: brownish grey then reddish brown or ochraceous brown.

Stem: cylindrical or slightly enlarged at the base, brittle, hollow, minutely scurfy and/or furrowed.

Flesh: ochraceous.

Spore-print: cream.

Spores: very long, broadly ellipsoid, pale honey, smooth but for some small granules at each end, and 16–23 × 11–14 μm in size; eight contained in an elongate cylindrical ascus.

Habitat & Distribution: Infrequent in gardens, on river-banks, sites of bonfires, etc., in spring.

Illustrations: F7[c]; LH 41; NB 41[3].

Plate 68. Morels and related fungi

All fruit-bodies at same magnification

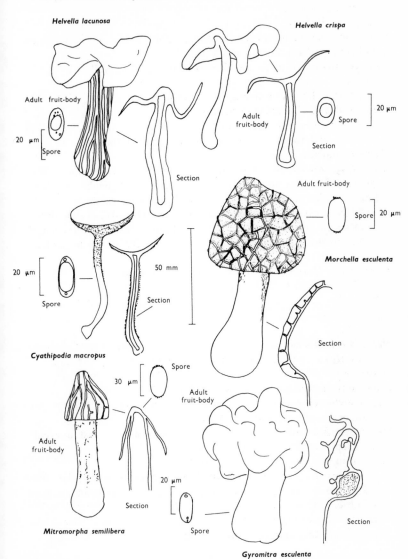

Helvella lacunosa

Adult fruit-body

20 μm
Spore

Section

Helvella crispa

Adult fruit-body

Spore

20 μm

Section

Adult fruit-body

Spore

20 μm

Morchella esculenta

Cyathipodia macropus

20 μm

Spore

Section

50 mm

Section

Spore

30 μm

Adult fruit-body

Section

20 μm

Spore

Section

Adult fruit-body

Mitromorpha semilibera

Gyromitra esculenta

Gyromitra esculenta (Persoon) Fries   Turban-fungus
Cap: width 30–40 mm; length 35–45 mm. *Stem:* width 15–25 mm; length 50–80 mm.
Description: Plate 68.
Fruit-body: consisting of a subglobose, more 'or less lobed, wrinkled and convoluted head on a short stem.
Cap: yellow-brown to reddish brown and becoming hollow or chambered.
Stem: flesh-coloured or creamy grey and powdery.
Flesh: yellow-buff, darker in the cap.
Spores: very long, ellipsoid, usually containing two or more yellowish oil drops and 18–22 × 9–12 μm in size; eight contained in an elongate cylindrical ascus.
Habitat & Distribution: This fungus is found in the spring, under conifers, but also on railway embankments, river banks, etc. This fungus is also known as the 'Lorel' or 'Elephant's ears'.
General Information: Mitromorpha semilibera (Fries) Léveille differs from species of *Morchella* in that the head is for its greater length free from the stalk. It is frequently found in the spring in gardens, tennis courts, etc.
Illustrations: G. esculenta—F 6^d; Hvass 327; LH 39.

Helvella crispa Fries   Common white helvella
Cap: width 18–28 mm. *Stem:* width 8–12 mm; length 40–65 mm.
Description:
Fruit-body: consisting of a saddle-shaped cap on a short stem.
Cap: convoluted towards the centre, two lobed, wavy at the margin, white or cream-coloured.
Stem: cylindric and hollow, white or cream-coloured and unevenly and deeply longitudinally furrowed.
Flesh: thin and pale.
Spores: very long, broadly ellipsoid, with a large central oil drop and 18–20 × 10–13 μm in size; eight contained in an elongate cylindrical ascus.
Habitat & Distribution: Frequent in damp woods with deciduous trees, from early summer until autumn.

General Information: Plate 68.

Helvella lacunosa Fries, 'Slate grey Helvella' is similar in stature but differs in being ash-grey or dark grey.

Leptopodia elastica (St Amans) Boudier is better placed in the genus *Helvella*. It differs in having a slender, smooth, cylindric stem and irregularly 2–3 lobed, yellow or tan-coloured cap.

Cyathipodia macropus (Fries) Dennis is sometimes placed in *Helvella*. It differs in having a grey cup-shaped cap on a long, slender stem. The spore-bearing tissue in the last species is the inner surface of the cup whilst in *Helvella* and *Leptopodia* it is on the outer surface of the saddle-like cap.

Mitrula paludosa Fries, the 'Bog beacon', is a similar fungus growing in spring to early autumn on old leaves and detritus in swamps. It is widespread and has a bright orange head on a white stem—as the common name might suggest. It grows to a height of about 20 mm.

Illustrations: H. crispa—F 6[e]; Hvass 331; LH 41. *H. lacunosa*—F 6[b]; Hvass 330; LH 39; NB 153[1]. *L. elastica*—Hvass 332; LH 39. *C. macropus*—F 6[a]; LH 39.

Rhizina undulata   Pine fire fungus

Fruit-body: width 20–60 mm, or several coalescing.

Description: Plate 69.

Fruit-body: chestnut-brown to rust colour with a distinct white or cream margin, fleshy, smooth, concave and thrown up into irregular humps.

Stem: lacking, but undersurface pale, ochraceous, and bearing numerous cylindrical branched, whitish root-like structures, 1–2 mm thick.

Flesh: reddish brown, tough and fibrous.

Spores: very, very long, spindle-shaped with two or more internal droplets, with hyaline extensions at each end, and 22–40 × 8–11 μm in size; eight contained in an elongate cylindrical ascus.

Habitat & Distribution: Infrequent in pine woods but common at the sites of bonfires in pine woodlands.

Daldinia concentrica (Fries) Cesati & de Notaris	Cramp-balls
Fruit-body: diameter 20–40 mm × 20–60 mm.
Description:

Fruit-body: date-brown at first finally black or dark brownish black, tough, minutely pimply over entire surface although at first covered in a powdery dust of asexual spores (conidia).

Stem: lacking.

Flesh: pale grey or buff, concentrically zoned with darker purplish black layers below which are small, black dots.

Spores: very long, black, ellipsoid with one flattened side and 12–17 × 6–9 μm in size; eight contained in an elongate cylindrical ascus.

Habitat & Distribution: Common on old deciduous wood, particularly of ash and beech.

General Information: These two fungi are unrelated; the first is related to the disc-fungi, like species of *Peziza*, whilst *Daldinia* is related to the Dead man's finger fungus. *Rhizina undulata* has been shown to be able to attack roots of pine or larch trees and cause death. *Daldinia* is a pure saprophyte rotting down wood into more simple compounds later to be incorporated into the soil-system. The common name 'Cramp-balls' refers to the old belief that if one of the fruit-bodies is carried in the pocket it saves the possessor from cramp and rheumatism. The other common name for the same fungus is 'King Alfred's cakes'. The black colour of the fruit-body is like that of charred cakes—resembling the cakes in the legend which King Alfred allowed to burn.

Illustrations: R. undulata—LH 37; NB 111[6]. *D. concentrica*—F 7[b]; LH 47; NB 147[7].

Xylosphaera polymorpha (Mérat) Dumortier	Dead man's fingers
Fruit-body: width 10–20 mm; length 30–60 mm.
Description:

Fruit-body: more or less club-shaped, irregularly or evenly lobed at apex, at first light brown due to development of asexually produced spores (conidia) but finally almost black.

Stem: black and short.

Flesh: white, fibrous and tough.

Plate 69. Cup-fungi allies

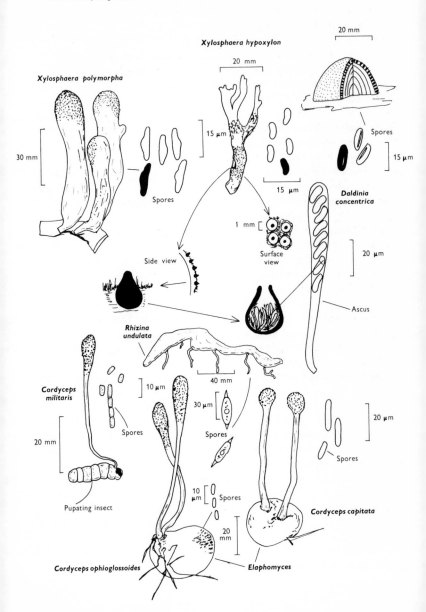

Xylosphaera polymorpha

Xylosphaera hypoxylon

20 mm

30 mm

Spores

15 μm

15 μm

Spores

15 μm

Daldinia concentrica

1 mm

Side view

Surface view

20 μm

Ascus

Rhizina undulata

40 mm

Cordyceps militaris

10 μm

30 μm

20 μm

Spores

Spores

Spores

20 mm

Pupating insect

10 μm

Spores

20 mm

Cordyceps capitata

Cordyceps ophioglossoides

Elaphomyces

Crust: black, thin, pimply with the protruding tips of the perithecia, and sometimes irregularly furrowed.

Spores: very long, fusiform with one flattened side, black and $20-32 \times 5-9$ μm in size; eight contained in an elongate cylindrical ascus.

Habitat & Distribution: Common either solitary or in clusters on dead stumps or on buried wood, especially that of beech. This fungus may be found throughout the year.

Xylosphaera hypoxylon Dumortier   Stag's horn fungus
Fruit-body: width 4–8 mm; length 25–60 mm.
Description: Plate 69.

Fruit-body: slender, subcylindrical to strap-shaped and usually forked repeatedly near the tip, white at first due to production of conidia and then black or dark brown and covered in pimples.

Stem: black and hairy.

Spores: very long, bean-shaped, black and $11-14 \times 5-6$ μm in size; eight in an elongate ascus.

General Information: Another name for *X. hypoxylon* is 'Candle-snuff fungus'. Other club-shaped ascomycetes include members of the genus *Geoglossum* (already mentioned p. 172) and members of the genus *Cordyceps*. Plate 69.

Cordyceps militaris (St Amans) Link, the 'Scarlet caterpillar fungus', produces orange-red or orange, minutely roughened fruit-bodies up to 50 mm high, which grow on larvae and pupae of moths buried in the soil. It is not infrequent late in the autumn in pasture land.

C. ophioglossoides (Fries) Link produces long (up to 100 mm high) yellow stemmed, dark and rough headed fruit-bodies growing on the subterranean fungus *Elaphomyces*—see p. 244.

C. capitata (Fries) Link also grows on fungi beneath the soil surface but has a rounded head. *Leotia lubrica* Persoon the 'Gum-drop fungus' is similarly coloured but grows on soil under trees and is gelatinous. It grows in autumn and is quite common and in fact more related to the Discomycetes than to *Cordyceps*.

Illustrations: X. *polymorpha*—F 7[d]; LH 47; NB 147[6]. X. *hypoxylon*—F 7[e]; LH 47; NB 147[5]. C. *militaris*—LH 48.

F. SPECIALISED HABITATS

Fungi of dung and straw heaps

Bolbitius vitellinus (Fries) Fries   Yellow cow-pat toadstool
Cap: width 20–40 mm. *Stem:* width 2–5 mm; length 30–60 mm.
Description:
Cap: chrome-yellow or lemon-yellow when young, paling with age
    at margin to become cinnamon-buff, bell-shaped but rapidly ex-
    panding to become plane or slightly umbonate, smooth, viscid but
    soon drying; margin striate then radially grooved, often split and
    the whole cap soon collapsing.
Stem: slender, whitish, cream colour to pale yellow, at apex covered
    in small, white floccose scales but downy at the base, fragile and
    soon collapsing.
Gills: adnexed or free, cinnamon-buff, thin and crowded.
Flesh: yellowish, very thin and lacking distinct smell.
Spore print: rust-brown.
Spores: long, yellow-brown under the microscope, ellipsoid with a
    very distinct germ-pore about 13 × 8 μm in size (11–15 × 6–9 μm).
Facial cystidia: rare, balloon-shaped.
Marginal cystidia: swollen, flask-shaped with a variable, elongate neck.
General Information: This fungus is common on horse droppings or
    other manures, but it may also be found amongst grass in pastures
    and in sand-dunes, and gardens on piles of rotting grass stems or
    straw. It is easily recognised by the colour and rapid expansion of
    the cap and the sudden collapse of the whole fruit-body. 'Vitellinus'
    means yolk of an egg and refers to the persistently bright yellow
    cap-centre, so obvious even when the fruit-body collapses. This
    collapsing is not one of autodigestion as described for members
    of the genus *Coprinus*. It is variable both in size and habitat, and I
    even have records of the fungus growing within herbaceous stems.
Illustrations: LH 153; WD 80[6].

Stropharia semiglobata (Fries) Quélet   Dung-roundhead
Cap: width 10–35 mm. *Stem:* width 4–7 mm; length 25–50 mm.
Description:

Cap: hemispherical or slightly umbonate, sometimes flattened and
   hardly expanding even with age, very viscid, smooth, pale yellow-
   ochre or yellowish tan.

Stem: slender, straight, white then yellowish, smooth, viscid, but then
   dry and shiny below an imperfectly formed, thin ring.

Gills: adnate, almost triangular in shape, crowded, dark brown to
   purplish black, but with ochraceous areas at maturity.

Flesh: pale ochre.

Spore-print: purplish brown.

Spores: very long, dark brown under the microscope, smooth, ellipsoid
   with large germ-pore and about 18 × 10 μm in size (17–20 × 9–10
   μm).

Facial cystidia: spindle-shaped, thin-walled and filled with amorphous
   contents which become yellow in solutions containing ammonia.

Marginal cystidia: spindle-shaped or flask-shaped, numerous, thin-
   walled and typically yellowing as above.

General Information: 'Semiglobata' means hemispherical and refers to
   the shape of the cap of *S. semiglobata*; it is a very variable fungus
   in both size of the cap and the prominence of the ring. The Dung-
   roundhead grows only on dung which is acidic in its soil status,
   whilst *Panaeolus semiovatus* (Fries) Lundell next described (p. 210)
   grows on slightly to distinctly base-rich dung. This may explain
   why in Britain the Dung-roundhead is the commoner of the two
   species. However, *P. semiovatus* was formerly placed in the genus
   Stropharia because of its blackish spores and distinct ring. The
   spores of *Stropharia* in the mass are violaceous black whilst those of
   P. semiovatus are brownish black. Under the microscope they are
   also differently coloured and have different chemical compositions
   as their reaction with dilute solutions of ammonia shows; the spores
   of the first species turn purplish olive in ammonia and those of the
   second species become very dark brown.

Illustrations: F 33[b]; Hvass 171; LH 153; NB 31[5]; WD 75[3].

Plate 70. Dung-fungi

Bolbitius vitellinus   **Stropharia semiglobata**

Panaeolus semiovatus   **Panaeolus sphinctrinus**

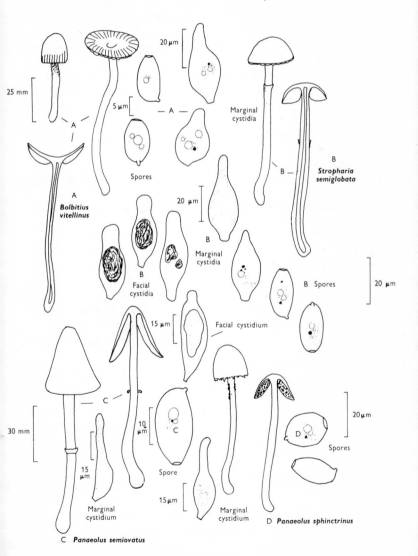

25 mm

20 μm

5 μm

— A —

Marginal cystidia

Spores

A
Bolbitius vitellinus

B
Stropharia semiglobata

20 μm

B
Marginal cystidia

B
Facial cystidia

B Spores

20 μm

15 μm

Facial cystidium

C

30 mm

10 μm

C

Spore

15 μm

Marginal cystidium

C **Panaeolus semiovatus**

15 μm

Marginal cystidium

D

Spores

20 μm

D **Panaeolus sphinctrinus**

Mottle-gills—on dung from Spring until Autumn.

Panaeolus semiovatus (Fries) Lundell

Cap: width 20–70 mm. *Stem:* width 5–10 mm; length 80–160 mm.
Description:

Cap: oval or bell-shaped, not expanding, dingy whitish or pale clay
   colour, smooth, slimy when moist, but soon drying and then becoming
   shiny, often wrinkled and cracked, and ornamented with fragments
   of veil at the margin.

Stem: dull, straight, rather rigid, tapering upwards, white, and striate
   at apex above a whitish erect and membranous, often collapsing, ring;
   yellowish below the ring and whitish and cottony at the slightly
   swollen base.

Gills: adnate, greyish then black, mottled and crowded.

Flesh: whitish or pale ochre.

Spore-print: black.

Spores: very long, very dark brown under the microscope with large
   obvious germ-pore and 18×10 μm (16–20×9–11 μm) in size.

Facial cystidia: flask-shaped and with amorphous contents.

Marginal cystidia: numerous, flask-shaped.

Panaeolus sphinctrinus (Fries) Quélet

Cap: width 15–35 mm. *Stem:* width 3–6 mm; length 60–95 mm.
Description:

Cap: bell-shaped, hardly expanding, expallent, dark grey to olivaceous
   black, much paler when dry and zoned when half dry; margin
   ornamented with a white fringe of veil fragments.

Stem: long, slender, straight, rather rigid but fragile, grey and completely powdered with white.

Gills: adnate, crowded and grey then blackened, mottled throughout
   except at the white fringed edge.

Flesh: reddish brown.

Spore-print: black.

Spores: long, very dark brown under the microscope, broadly lemon-
   shaped with large germ-pore, smooth and 14–15×9–10 μm in size
   (14–19×8–10×10–12 μm).

Facial cystidia: absent.

Marginal cystidia: numerous, cylindrical, flexuous and hyaline.

General Information: P. sphinctrinus is recognised by the overall grey
   colouration and very distinct white fringe to the cap-margin.

   P. campanulatus (Fries) Quélet which is said to be common is in fact infrequent and many records really refer to *P. sphinctrinus*. The word *semiovatus* means half ovate and refers to the shape of the cap in *P. semiovatus*. *Sphinctrinus* means banded, referring to the zoned cap of the fungus when it is partially dry.

Illustrations: P. semiovatus—LH 145; WD 77³. *P. sphinctrinus*—NB 41⁵; WD 78¹.

Coprinus cinereus (Fries) S. F. Gray   Dung-heap ink-cap

Cap: width 20–40 mm. *Stem:* width 4–8 mm; length 50–100 mm.

Description:

Cap: oval then rapidly expanding, covered at first in a mass of dense, white or greyish woolly scales which break up into patches and finally leave the cap shiny, brownish grey at centre and striate and dark grey at the margin.

Stem: white, covered particularly towards the base with white, woolly scales, long, fragile, tapering upwards and at the base often elongated into a 'tap root' buried in the dung.

Gills: free, white but then rapidly dissolving into a black liquid.

Flesh: thin and whitish.

Spore-print: violaceous black.

Spores: medium sized, ellipsoid, smooth with a distinct germ-pore and 10–12 × 5–6 μm in size.

Facial cystidia: absent.

Marginal cystidia: inflated and large.

General Information: It is found on manure heaps, on straw dung and on silage heaps: very common throughout the year.

C. macrocephalus (Berkeley) Berkeley is very closely related to *C. cinereus*, but differs in having much larger spores over 12–15 × 7–9 μm, a long cap and a stem which lacks a rooting base.

Coprinus radiatus (Fries) S. F. Gray is smaller in stature and also differs in spore-size (11–14 × 6–7 μm). *C. pseudoradiatus* Kühner & Josserand is minute and has even smaller spores (7–9 × 4–5 μm). The dung-heap ink-cap has long been used by scientists in genetic studies, usually under the name of *C. lagopus* (Fries) Fries. However, this latter species, although similar, grows only on woodland detritus; it has narrower spores. The dung-heap ink-cap may be referred to in other books as *C. fimetarius* Fries or *C. macrorhizus* (Fries) Rea and whilst *cinereus* means grey referring to the colour, *fimetarius* means dung—from the habitat, and *macrorhizus* refers to the long rooting base found in some specimens.

Illustrations: LH 137; NB 41¹⁰; WD 81⁴.

The genus *Coprinus*—or Ink-caps

The genus *Coprinus* is easily recognised from all other agarics by the structure and development of the fruit-body. In the field, most species of the genus can be recognised by the gradual conversion of the gills, and often the cap tissue into a black liquid resembling ink—hence the name inky-caps. The conversion of the gills to an inky mass is called autodigestion and the process is complete within a few hours; this mechanism enables spores to be dispersed immediately they have ripened. Unlike other agarics the spores are not shot off into the spaces between the gills, but directly into the air. The gills are parallel-sided in *Coprinus* and not wedge-shaped as in more normal agarics, and in order to achieve spore dispersal the gills must disintegrate; Coprini are very specialised.

Coprinus is a large genus with over seventy members in the British Isles, many of which are strictly dung-loving. It is impossible to give more than one example in full here, for although many of the large species can be recognised on sight the smaller ones require the aid of a microscope. The interested student must therefore refer to more advanced texts, but in order to demonstrate the diversity of the Coprini and how they are classified the following key to the sections of *Coprinus* will be found useful.

1. Cap naked of any veil fragments, either smooth or covered in minute hairs 2
   Cap covered when young by powdery or hairy veil, particles of which either may persist on the cap until maturity or may disappear quickly 3
2. Cap completely naked—group Nudi, e.g. *C. miser* (Karsten) Karsten
   Cap with hairs giving it a frosted appearance—group Setulosi, e.g. *C. ephemerus* (Fries) Fries, *C. pellucidus* Karsten and *C. bisporus* J. Lange
3. Veil on the cap composed under the microscope of rounded cells giving the cap a floccose powdery appearance—group Vestiti, e.g. *C. patoullardii* Quélet, *C. niveus* (Fries) Fries and *C. ephemeroides* (Fries) Fries
   Veil on the cap composed under the microscope of elongate cells, either like thin-hairs or strings of sausages 4

Plate 71. Dung-fungi — The genus *Coprinus*

Microscopic characters variously magnified

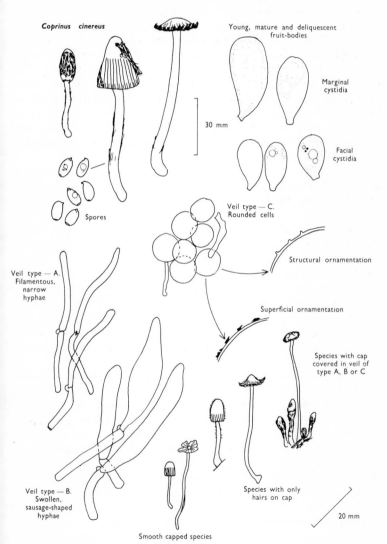

Coprinus cinereus

30 mm

Spores

Young, mature and deliquescent fruit-bodies

Marginal cystidia

Facial cystidia

Veil type — C. Rounded cells

Structural ornamentation

Veil type — A. Filamentous, narrow hyphae

Superficial ornamentation

Species with cap covered in veil of type A, B or C

Veil type — B. Swollen, sausage-shaped hyphae

Species with only hairs on cap

20 mm

Smooth capped species

4. Veil on the cap composed under the microscope of strings of sausage-shaped cells—group Lanatuli, e.g. *C. cinereus*, *C. pseudoradiatus*, *C. radiatus* (see p. 211)

Veil on the cap composed under the microscope of thick- or thin-walled, flexuous or straight, filamentous, hardly inflated cells—group Impexi, e.g. *C. filamentifer* Kühner, *C. vermiculifer* Dennis.

General notes on dung-loving fungi and their habitats

Dung fungi are highly satisfactory for demonstrating the diversity and morphology of a group of related organisms within a single ecological system, as representatives of most of the major groups of fungi usually grow on dung after a period of incubation. Dung will always produce characteristic fungi whatever time of year it is collected.

Dung is best incubated in a light place, for example on a window sill, in a warm room on layers of blotting paper or other absorbent material. For rabbit-pellets and samples of similar size petri-dishes are ideal, but for cow, horse and similar types of dung large covered dishes such as casseroles or sandwich containers are very good. Samples should not be kept in airtight containers for long periods of time as under such conditions animal life present rapidly breaks down the dung and induces anaerobic conditions. Instead larvae and earthworms should be excluded from the sample as they decompose the dung and inhibit fungal growth but their activity can be reduced, if causing a problem, by spraying the sample lightly with a proprietary fly-kill aerosol.

By keeping the dung under constant observation during incubation a whole succession of fungi can be seen. Thus the first fungi to appear are the moulds which although numerous need a microscope for their identification. The moulds are followed by a series of Ascomycetes (*Sporormia* & *Sordaria* with flask-shaped fruit-bodies and *Iodophanus*, *Coprobia* and *Cheilymenia* with disc-shaped fruit-bodies), which are best sought with the use of a powerful hand-lens or a stereoscopic binocular miscroscope when their full beauty will be revealed. However, because they need the aid of instruments even to see them they cannot be considered larger fungi. The fruit-bodies of the Basidiomycetes are readily seen with the naked eye, but a hand-lens is still very useful

Plate 72. Dung-fungi: Cup fungi and allies

Magnification various; Fruit-bodies as seen under a dissecting microscope; Microscopic characters under dry high power microscope

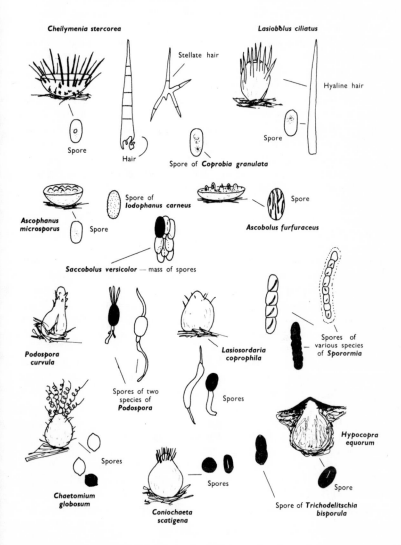

Cheilymenia stercorea

Stellate hair

Spore

Hair

Spore of **Coprobia granulata**

Lasiobolus ciliatus

Hyaline hair

Spore

Spore of **Iodophanus carneus**

Ascophanus microsporus

Spore

Ascobolus furfuraceus

Spore

Saccobolus versicolor — mass of spores

Podospora curvula

Spores of two species of **Podospora**

Lasiosordaria coprophila

Spores

Spores of various species of **Sporormia**

Chaetomium globosum

Spores

Coniochaeta scatigena

Spores

Hypocopra equorum

Spore

Spore of **Trichodelitschia bisporula**

for observing features of the cap and stem, particularly the veil characters. The Basidiomycetes usually conclude the succession of fungi
found on dung and soon after this state the dung is colonised by mosses
and higher plants and later it is fully incorporated into the soil.

Dung is a very useful substrate for studying succession. However,
equally interesting results can be obtained from observing the fungi
which appear on a stump, colonise a newly laid lawn, or indeed those
growing on refuse such as a cast-out rug; microscopic and larger fungi
are all to be found.

If the dung cannot be incubated immediately it should be dried
quickly, for most dung-fungi will survive such treatment and grow
when the sample is remoistened. The blotting-paper on which the
dung is placed should be kept moist throughout the incubation period.

One large discomycete (up to 80 mm across) occurring on manure-
heaps must, however, be mentioned, this is *Peziza vesiculosa* St Amans
(see p. 200); the inner surface of this cup-fungus becomes detached
from the flesh at maturity and forms blisters.

(ii) Fungi of bonfire-sites

Pholiota highlandensis (Peck) A. H. Smith   Charcoal pholiota
Cap: width 20–50 mm. *Stem:* width 4–8 mm; length 25–60 mm.
Description: Plate 73.
Cap: convex then flattened and slightly umbonate, smooth, very
    sticky at first, but becoming shiny when dry, orange-yellow to sand-
    colour; the margin is first incurved and ornamented with filaments
    from the veil, but these are soon lost.
Stem: dirty yellow, darker towards the base, cylindric or narrowed
    downwards and covered in small fibrillose scales.
Gills: clay-coloured then dull brown, adnate and crowded.
Flesh: yellowish.
Spore-print: dull rust-brown.
Spores: medium-sized, ellipsoid, smooth, dull brown under the micro-
    scope and 7–8 × 3–4 μm in size.
Facial cystidia: spindle-shaped with obtuse apex.
Marginal cystidia: similar to facial cystidia but usually smaller.
General Information: This fungus which occurs from spring to autumn
    is recognised by the habitat, colour of the fruit-body and the spore-

size. It is known in many books as *Flammula carbonaria* (Fries) Kummer, but the genus *Flammula* is no longer used for it refers to a flowering plant in the buttercup family.

P. *highlandensis* is the same fungus as that referred to as *Pholiota carbonaria* by European Mycologists, but this name cannot be used for it refers to an entirely different N. American species. 'Highlandensis', in fact, refers to the locality where the present fungus was first found in the United States of America. The true *P. carbonaria* A. H. Smith has only been found once in Europe and this only recently in the south of England. It differs in the reddish orange scales on the stem; indeed it is a much brighter fungus than the common charcoal *Pholiota*.

Tephrocybe anthracophila (Lasch) P. D. Orton
Cap: width 1–4 mm. *Stem:* width 1 mm; length 2–5 mm.
Description: Plate 73.
Cap: blackish when wet, drying sooty brown, slightly depressed in the centre, smooth, and viscid.
Stem: sooty brown, tough and smooth.
Flesh: sooty brown.
Gills: whitish then grey, adnate and not very crowded.
Spores: medium sized, subglobose, 4–6 × 4–5 μm in diameter and minutely roughened.
Spore-print: white, not blueing in solutions of iodine.
General Information: T. *atrata* also grows on burnt soil and is very closely related, but differs in its spores being broadly ellipsoid and smooth. *Mycena leucogala* also grows on burnt soil (see p. 88).
Illustrations: T. *anthracophila*—LH 83. T. *atrata*—WD 4[b].

Psathyrella pennata (Fries) Pearson & Dennis   Bonfire brittle-cap
Cap: width 10–30 mm. *Stem:* width 1–3 mm; length 30–40 mm.
Description:
Cap: conical or bell-shaped then expanding and slightly umbonate, whitish because of a coating of dense fibrils, but soon becoming brownish as these are lost.
Stem: short, stout, white and densely floccose.
Gills: slightly adnate, pale brownish grey with pink tinge, then dark-brown.
Spore-print: purplish brown.
Spores: medium sized, oval, ellipsoid with an obvious germ-pore, purplish brown under the microscope and 8–9 × 4–5 μm in size.
Marginal & facial cystidia: flask-shaped, hyaline with either a short or long neck.
The brown-spored *Hebeloma anthracophila* Maire is similar.

Coprinus angulatus Peck   Bonfire ink-cap
Cap: width 4–25 mm. *Stem:* 1–3 mm; length 15–30 mm.
Description:
Cap: dark red-brown at first, then orange-brown, especially at the margin and appearing as if frosted all over, conical at first but rapidly expanding at the margin and becoming grey-brown, strongly striate and deliquescent, leaving finally only a central red-brown umbo.
Stem: white and minutely hairy.
Gills: free, dirty whitish then black.
Spore-print: black-brown.
Spores: medium sized, dark brown under the microscope, lobed like the hat of a bishop and 8–11 × 6–8 × 5–7 μm in size.
Marginal cystidia: bottle-shaped, very variable.
Facial cystidia: similar to marginal cystidia.
General Information: It must be noted that this fungus has spores which require three quite different measurements to describe the dimension. Another species of *Coprinus* found on burnt soil is *C. lagopides* Karsten which resembles *C. cinereus* (Fries) S. F. Gray (p. 211); it is typified, however, by the rounded spores.

Plate 73. Fungi of bonfire-sites

Pholiota highlandensis   **Coprinus angulatus**   **Psathyrella pennata**   **Hebeloma anthracophila**

Tephrocybe anthracophila

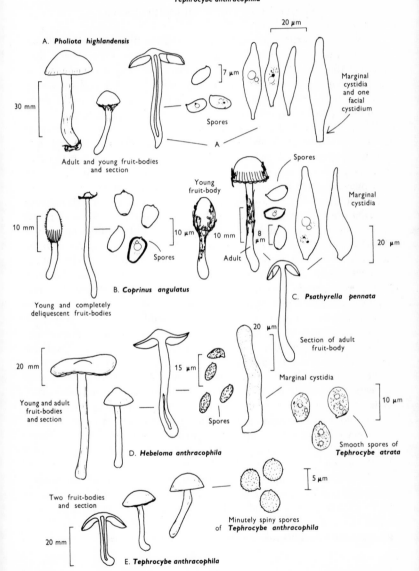

A. *Pholiota highlandensis*

20 µm

7 µm

Spores

Marginal cystidia and one facial cystidium

A

30 mm

Adult and young fruit-bodies and section

10 mm

10 µm

Spores

B. *Coprinus angulatus*

Young and completely deliquescent fruit-bodies

Young fruit-body

Spores

10 mm

10 µm

Adult

8 µm

Marginal cystidia

20 µm

C. *Psathyrella pennata*

20 mm

Young and adult fruit-bodies and section

15 µm

Spores

20 µm

Section of adult fruit-body

Marginal cystidia

10 µm

Smooth spores of *Tephrocybe atrata*

D. *Hebeloma anthracophila*

Two fruit-bodies and section

20 mm

5 µm

Minutely spiny spores of *Tephrocybe anthracophila*

E. *Tephrocybe anthracophila*

General notes on fungi of burnt sites

Several common fungi found at the sites of bonfires have their closest relatives amongst various groups of microscopic fungi more than amongst the large forms already discussed. Keeping a close watch at the site of a former bonfire day by day, week by week and month by month is very rewarding and shows a further example, like the dung habitat, of a tightly knit community of various groups of fungi.

Peziza repanda Persoon has been discussed in detail above (p. 200); its close relatives *P. petersii* Berkeley & Curtis (brown with grey tints and with spores finely warted and measuring $10–12 \times 5–6$ μm), *P. praetervisa* Bresadola (violet or mauve and with spores finely warted and measuring $11–13 \times 6–8$ μm), *P. violacea* Persoon (dark violet with smooth spores measuring $13–15 \times 7–9$ μm) and *P. echinospora* Karsten (dark chocolate brown with spores densely warted and $14–18 \times 7–10$ μm in size) all grow on the sites of old bonfires or around charred root stumps. *Rhizina undulata* also found by charred stumps has been described on p. 203. These are large to medium sized disc-fungi, but there are many much smaller species which cannot be dealt with here, such as species of *Anthracobia* and *Trichophaea*. Pyreno-mycetes are also found on charred wood and soil. Probably the commonest species of fungus met with is a pale reddish orange to rose-pink disc-fungus seated on a white mycelial mat; this is *Pyronema omphalodes* (St Amans) Fuckel. *Morchella esculenta* St Amans and *M. elata* Fries (see p. 200) appear to grow on the sites of garden bonfires or where cinders have been spread on the soil surface. The stimulus for fruiting appears to be due to the release of mineral nutrients during the process of burning. Competition from other fungi appears to be reduced so rapid colonisation by the bonfire fungi (carbonicoles) after the period of sterilisation ensures their development. Many similar fungi were found about bomb- and shell-craters on the continent during the two World Wars.

One microscope fungus, however, must be mentioned when considering bonfires and that is *Neurospora sitophila* Shear & Dodge so much used in genetical studies. It can be found as the conidial state on burnt soil and is called 'Baker's mould' because it is frequently found growing on refuse in the hot moist conditions of bakers' kitchens.

Plate 74. Fungi of bonfire-sites

Magnification various; Fruit-bodies as seen under a dissecting microscope
Microscopic characters under dry high power of microscope

Cup fungi and allies

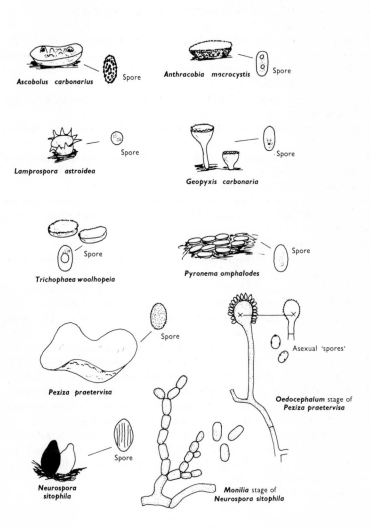

Ascobolus carbonarius   Spore

Anthracobia macrocystis   Spore

Lamprospora astroidea   Spore

Geopyxis carbonaria   Spore

Trichophaea woolhopeia   Spore

Pyronema omphalodes   Spore

Peziza praetervisa   Spore

Oedocephalum stage of **Peziza praetervisa**   Asexual 'spores'

Neurospora sitophila   Spore

Monilia stage of **Neurospora sitophila**

(iii) Fungi of bogs and marshes

Sphagnum **bogs**
Hypholoma elongatum (Fries) Ricken
Cap: width 12–20 mm. *Stem:* width 3–5 mm; length 50–80 mm.
Description: Plate 75.
Cap: bell-shaped but rapidly expanding to become plane, honey-yellow with a greyish green tint, slightly striate at the margin and also with a few remnants of a fibrillose veil when very young, but these are soon lost.
Stem: slender, smooth, whitish at the apex and yellow-brown or honey-yellow below.
Gills: adnate and distant, pale ochraceous honey-yellow then lilaceous grey and finally sepia.
Flesh: yellowish in the cap, red-brown in the stem and lacking a distinct smell.
Spore-print: purplish brown.
Spores: long, ellipsoid, fairly thick-walled, olivaceous brown under the microscope and with a small germ-pore, smooth and 10–12 × 6–7 μm in size.
Marginal cystidia: flask-shaped and hyaline.
Facial cystidia: flask-shaped with contents which turn yellowish in solutions containing ammonia.
General Information: This fungus which appears from early summer to late autumn is recognised by the almost uniform ochraceous colour with hint of olive and its habit of growing in troops. The word *elongatum* means elongated and refers to the shape of the stem which pushes up through the *Sphagnum* and in order to disperse its spores it must elongate so that it just pushes up above the bog-surface. *H. polytrichi* is closely related to *H. elongatum* but has a paler cap and stem and it grows in moss, particularly *Polytrichum* in woodlands; the spores of *H. polytrichi* are paler, slightly narrower and slightly thinner, but they have a much more distinct germ-pore.
  Both the above species have been formerly placed in *Psilocybe*, but they are more correctly classified in *Hypholoma* along with the sulphur-tuft fungus (see p. 64) because of the cortina-like veil and specialised facial cystidia.
Illustrations: WD 78[5].

Tephrocybe palustris (Peck) Donk

Cap: width 12–30 mm. *Stem:* width 3–5 mm; length 50–75 mm.
Description: Plate 75.

Cap: bell-shaped then plano-convex, but finally depressed at centre, watery buff to greyish with flush of ochre or smoky grey, striate to centre when moist, but drying out non-striate and uniformly ochraceous buff.

Stem: thin, rather long, smooth, similarly coloured to the cap or paler, fragile and whitish woolly at the base.

Gills: dirty whitish, adnate with a tooth and not very crowded.

Flesh: thin, watery buff, drying out ochraceous and with a strong smell of new meal.

Spore-print: white.

Spores: medium sized, hyaline under the microscope, oval, not turning blue-grey in solutions of iodine, and 6–7 × 4–5 μm in size.

Marginal and facial cystidia: absent.

General Information: This fungus which grows from late spring to autumn is usually associated with a greying and finally a killing of the *Sphagnum*, noticeable from a distance even in the absence of the fruiting-bodies as paler patches in the rich green bog. Another agaric found only in *Sphagnum* bogs is *Omphalina sphagnicola* (Berkeley) Moser with decurrent gills and long, elongate, hyaline spores.

At the margin of *Sphagnum* bogs, the fungus *Mycena bulbosa* can be found attached to the base of tufts of rushes.

Potting up a sward of *Sphagnum* and retaining it in a warm greenhouse during winter favours the bog agarics to fruit when other larger fungi are not available.

Mycena bulbosa (Cejp) Kühner

Cap: width 3–6 mm. *Stem:* width 1 mm; length 10–15 mm.
Description:

Cap: dirty white, greyish and very gelatinous.

Stem: very thin, hyaline with a very distinct hairy, basal disc.

Gills: crowded, adnexed, very short and whitish.

Spore-print: white, but because it is so small it is often difficult to see.

Spores: medium sized, hyaline under the microscope, ellipsoid, not blueing in solutions of iodine, and 8–10 × 4 μm in size.

Marginal cystidia: clavate or ventricose, hyaline and smooth.

Facial cystidia: absent.

Illustrations: T. *palustris* LH 83.

Galerina paludosa (Fries) Kühner

Cap: width 10–20 mm. *Stem:* width 3–5 mm; length 50–90 mm.

Description:

Cap: conico-convex expanding slightly but retaining the central umbo, striate to half-way, sand-colour to red-brown, hygrophanous, minutely floccose because of remnants of veil distributed over its surface, but soon becoming smooth.

Stem: long, buried amongst the *Sphagnum*, red-brown and flecked with white fibrils, except at the finely hairy apex, the fibrils typically form a distinct but easily lost ring.

Gills: almost horizontal, adnate to subdecurrent, pale at first and then rust-brown.

Spore-print: rust-brown.

Spores: medium-sized, ovate to slightly lemon-shaped, minutely warty, honey-brown under the microscope and about 10 × 6 μm in size, (9–11 × 6–7 μm).

Facial cystidia: absent.

Marginal cystidia: hyaline, almost cylindrical or bottle-shaped with an inflated base.

General Information: This species grows from spring to early autumn in *Sphagnum* bogs; several other species of *Galerina* are also found in the same localities:—

(i) *G. sphagnorum* (Fries) Kühner has a convex cap, fibrillose silky and ochraceous brown stem, but it lacks the ring-zone so typical of *G. paludosa*. The smell is like that of meal when crushed and the gills are emarginate.

(ii) *G. tibiicystis* (Atkinson) Kühner has a rapidly expanding cap which becomes plano-convex or depressed at maturity; it also lacks a ring-zone, but the stem in this species is finely hairy because of the presence of numerous pin-shaped cells which can be seen only with the aid of a lens. The gills are broadly adnate.

Illustrations: G. paludosa—LH 175.

Plate 75. Fungi of marshes

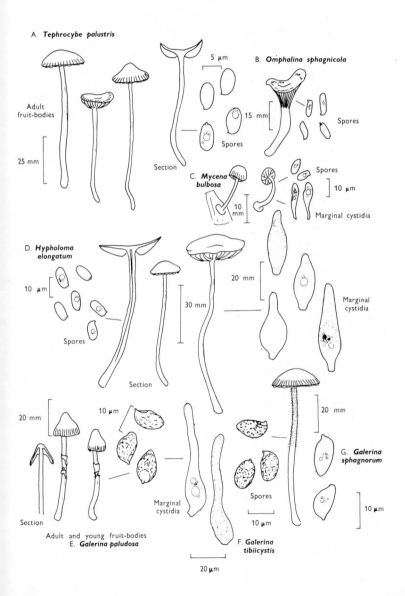

A. *Tephrocybe palustris*

Adult
fruit-bodies

25 mm

Section

5 µm

15 mm

Spores

B. *Omphalina sphagnicola*

Spores

C. *Mycena bulbosa*

10 mm

Spores

10 µm

Marginal cystidia

D. *Hypholoma elongatum*

10 µm

Spores

Section

20 mm

30 mm

Marginal cystidia

20 mm

10 µm

Section

Adult and young fruit-bodies
E. *Galerina paludosa*

Marginal cystidia

Spores

F. *Galerina tibiicystis*

10 µm

G. *Galerina sphagnorum*

10 µm

20 µm

(b) Alder-carrs

Naucoria escharoides (Fries) Kummer
Cap: width 12–30 mm. *Stem:* width 1–3 mm; length 25–45 mm.
Description: Plate 76.
Cap: pale yellowish ochre, but becoming darker ochraceous with age, scurfy, convex but then flattened, or with its edge upturned; the margin is slightly striate when moist.
Stem: slender, pale to dirty yellowish ochre but darker brown at base, slightly fibrillose, particularly at first because of filaments from a veil, but these are soon lost.
Gills: pale tan to brownish ochre with a paler, floccose margin, adnate and crowded.
Flesh: yellowish ochre but lacking a distinct smell.
Spore-print: clay-colour.
Spores: medium sized, almond-shaped, pale brown under the microscope, warted and 10–11 × 5–6 μm in size.
Marginal cystidia: swollen below, but drawn out into a hair-like apex.
Facial cystidia: absent.
General Information: Although this is a common species growing in damp places under alder it is difficult except with an expert eye to separate it from several closely related species which are also found in similar places. At present it is not known whether these fungi are favoured by the water-logged base-rich, reducing soils found nowhere else except under alder, or if they have a special relationship with the tree. There is ample evidence that soil conditions in alder woods are rather different from those found in other woodlands, but whatever the reason *Naucoria escharoides* is only found under alder—in fact this species has been placed in the genus *Alnicola* because of this character—*cola* meaning inhabitant and *Alnus* the tree of that name. Willow-carrs have not been as extensively studied as alder-carrs but there is evidence that a store of mycological information is still to be obtained from these places. Several species of *Naucoria* have been described from only willow-carrs, while others are to be found under both alder and willows; about eight species are known to grow under alder. The word *escharoides* means scab-like and refers to the cap which when freshly collected is minutely scaly and appears scabby.
Illustrations: LH 163; WD 67[1].

(iv) Fungi of beds of herbaceous plants

Beds of herbaceous plants provide protection for many small agarics and collecting can be conducted in these situations from spring to early winter. The buffered environments under the herbs is humid and relatively still, and this allows the development of the small often delicate fruit-bodies of certain species to continue unimpeded. Nettle-beds or mixtures of nettle and dog's mercury have very rich floras under the shelter of their leaves and stems, either on the bare soil or plant debris.

On herbaceous stems
Coprinus urticicola (Berkeley & Broome) Buller
Cap: width 4–7 mm. *Stem:* width 1 mm; length 10–15 mm.
Description: Plate 76.
Cap: white then greyish, globose at first and then expanding to become plane with upturned margin covered, at first, with scales from a veil which at the centre are white-tipped with ochre.
Stem: white and slightly downy.
Spore-print: brownish black.
Spores: elliptic-ovoid, only slightly compressed with distinct germ-pore, dark brown under the microscope and 6–8 × 5 μm in size.
Marginal cystidia: ellipsoid to pyriform and hyaline.
Facial cystidia: elongate cylindric larger than marginal cystidia.

On bare soil
Leptonia babingtonii (Bloxam) P. D. Orton
Cap: 5–15 mm. *Stem:* width 1 mm; length 20–50 mm.
Description: Plate 76.
Cap: grey to sepia or greyish brown entirely scaly-hairy, at first, but then fibrillose.
Stem: silvery grey to grey-sepia and silky fibrillose.
Gills: greyish pink.
Spore-print: greyish pink.
Spores: very long, wavy angular in outline, very pale honey under the microscope and 14–20 × 7–9 μm.
Marginal cystidia: club-shaped or balloon-shaped and hyaline.
Facial cystidia: absent.
So very different to other species of *Leptonia* is it that it should be classified in Dr. Pilát's genus *Pouzaromyces*.

Conocybe mairei Watling

Cap: width 5–10 mm. *Stem:* width 1 mm; length 10–40 mm.

Description:

Cap: pale to deep ochraceous or buff, minutely tomentose.

Stem: flexuous, whitish or very pale ochraceous.

Gills: pale buff then ochraceous.

Spore-print: ochraceous.

Spores: medium sized, ellipsoid or slightly almond-shaped with small germ-pore and 6–8 × 3–4 μm in size.

Flammulaster granulosa (J. Lange) Watling

Cap: 4–15 mm. *Stem:* width 1 mm; length 10–25 mm.

Description:

Cap: ochraceous to date-brown, darker at the centre and granular scaly throughout.

Stem: similarly coloured to the cap and similarly roughened, except for the slightly smoother paler apex.

Spores: ellipsoid to almond-shaped, very pale brown under the microscope and 8–10 × 4–5 μm in size.

Marginal cystidia: cylindric-wavy, hyaline.

Facial cystidia: absent.

Depending on the herbaceous constituents the fungus-flora will vary. Certain species are found on all sorts of herbaceous debris, but others are much more specific to their substrate preferences. Beds of Butterbur, Coltsfoot or Impatiens are also good hunting places, as are beds of sedges in fenland. In many of these localities agarics with reduced fruit-bodies looking like disc-fungi are frequently seen. We have already discussed the specific requirements of certain species of *Marasmius* (see p. 92).

Plate 76. Fungi of alder-carrs and from under herbaceous plants

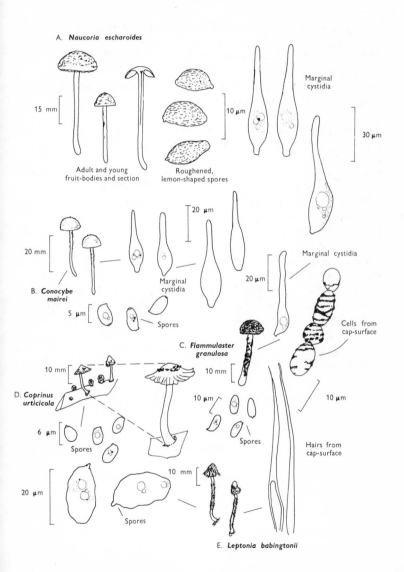

A. **Naucoria escharoides**

15 mm

Adult and young
fruit-bodies and section

Roughened,
lemon-shaped spores

10 μm

Marginal
cystidia

30 μm

20 mm

B. **Conocybe
mairei**

20 μm

Marginal
cystidia

5 μm

Spores

20 μm

Marginal cystidia

C. **Flammulaster
granulosa**

Cells from
cap-surface

10 mm

D. **Coprinus
urticicola**

10 mm

10 μm

Spores

6 μm

Spores

10 μm

Hairs from
cap-surface

10 mm

20 μm

Spores

E. **Leptonia babingtonii**

(v) Fungi of moss-cushions

Many small species grow amongst moss cushions on tree trunks, tucked in crevices in walls or on the tops of old buildings. However, there is one genus of agarics, i.e. *Galerina* which is probably more typical than any other of such situations. There are many members of this genus whose small caps are found in the autumn pushing up through the moss plants. Plate 78.

Galerina hypnorum (Fries) Kühner

Cap: width 4–6 mm. *Stem:* width 1 mm; length 20–40 mm.
Description:

Cap: hemispherical or bell-shaped, hygrophanous, orange-yellow, sand-colour, smooth and striate almost to the cap-centre.

Stem: smooth and similarly coloured to the cap.

Gills: yellow-tawny then rust-coloured, adnate emarginate, rather broad and somewhat distant.

Flesh: thin, yellow-tawny and with a smell of new meal.

Spore-print: rust-colour.

Spores: medium-sized, almond-shaped, golden yellow under the microscope, slightly roughened and $10–11 \times 6–7$ μm in size.

Marginal cystidia: flask-shaped or cylindrical with slight swelling at the apex.

Facial cystidia: absent.

Galerina mycenopsis (Fries) Kühner

Cap: width 6–15 mm. *Stem:* width 1 mm; length 30–60 mm.
Description:

Cap: similarly coloured to *G. hypnorum*, but with a few white silky fibrils.

Stem: coloured as the cap, but with white silky fibrils when young.

Gills and flesh: as in *G. hypnorum*, but it has no smell.

Spore-print: rust-colour.

Spores: medium-sized, ellipsoid, pale golden yellow under the microscope, smooth and $9–11 \times 5–6$ μm in size.

Marginal cystidia: club-shaped, cylindrical and with distinct rounded heads.

Facial cystidia: absent.

General Information: G. *mniophila* (Lasch) Kühner is similar to or slightly larger than *G. mycenopsis*, but differs in its dull honey-coloured cap and stem, and distinctly roughened spores. *G. calyptrata* P. D. Orton is small and has been long confused with *G. hypnorum*; it, however, is of a much brighter orange-colour, with distinct white fibrils on the cap and has spores which have a distinct envelope, sometimes separating as a loose covering. *G. vittaeformis* (Fries) Moser is a red-brown fungus with 2-spored basidia, facial cystidia, minutely hairy stem, and very rough spores; it grows in moss in pastures as well as on moss-cushions.

(vi) Heath and mountain fungi

Moorland fungi

Marasmius androsaceus (Fries) Fries   Horse-hair toadstool
Cap: width 5–15 mm. *Stem:* width 1 mm; length 30–60 mm.
Description: Plate 77.
Cap: whitish to pale smoke-brown with a distinct wine-coloured tinge, membranous, flattened, or umbilicate and radially wrinkled.
Stem: thread-like, black or very dark brown, horny and usually springing from a black horse-hair-like mycelium.
Gills: whitish or dirty flesh-colour, adnate and crowded.
Flesh: white in the pileus and black in the stem.
Spore-print: white.
Spores: medium-sized, pip-shaped, not blueing in solutions containing iodine and measuring $7–9 \times 3–4$ μm in size.
Marginal cystidia: oval or ellipsoid, covered on the upper half with small pimple-like projections.
Facial cystidia: absent.
General Information: This fungus is common in troops from late summer until winter on dead and dying heather. It is also found in woods on leaves and twigs, particularly in plantations on conifer needles. It is easily recognised by the dark horse-hair-like stem which becomes bent and twisted on drying and the small, pinkish flesh-coloured cap. The word *androsaceus* means, and refers to, the stem which resembles the tough and wiry fronds of some of the red algae, such as *Ahnfeldtia* which is found around our sea-shores.
Illustrations: LH 115; NB 47[1]; WD 24[4].

Omphalina ericetorum (Fries) M. Lange
Cap: width 5–20 mm. *Stem:* width 2 mm; length 10–20 mm.
Description:
Cap: variable in colour, straw-colour, cream-colour, bistre or grey, convex then flat or slightly depressed, radially grooved to the centre when moist; the margin is scalloped.

Stem: slender, similarly coloured to the cap, except for a brownish wine-coloured zone at the very apex, thickened upwards and smooth with a white and woolly base.

Gills: adnate to decurrent, white then cream-colour or yellowish, triangular in shape, very distant and often connected by veins.

Flesh: pale cream-colour.

Spore-print: white.

Spores: medium sized, hyaline under the microscope, broadly ellipsoid, or pip-shaped, not becoming bluish grey in solutions of iodine, 8–10 × 5 µm in size.

Marginal and facial cystidia: absent.

General Information: This fungus is common and often in large troops on peaty ground in woods as well as in moorland and mountain regions. In mountains *O. ericetorum* must be carefully distinguished from some of the truly mountain species of *Omphalina* dealt with on p. 236. *O. wynniae* (Berkeley & Broome) P. D. Orton is similar but pale lemon-yellow and is found on stumps of conifers. The word *ericetorum* refers to the habit of growing on heaths—*Erica* is the Latin name for heath. In many books this same fungus is called *O. umbellifera* which reflects the shape of the cap—umbrella shaped.

Illustrations: Hvass 116; LH 99; NB 85[7]; WD 29[9].

Entoloma helodes (Fries) Kummer
Cap: width 25–75 mm. *Stem:* width 2–6 mm; length 25–55 mm.
Description:
Cap: finely or minutely velvety at centre, fibrillose or white silky as if frosted towards the margin, sepia or bistre, or mouse-grey, dull-coloured but with a hint of violaceous brown.

Stem: equal or slightly thickened at the apex, sometimes club-shaped, thickened at the base, greyish brown and pale cream-colour at the base.

Flesh: dark sepia in the cap, whitish in the stem and smelling strongly of meal.

Plate 77. Moorland fungi

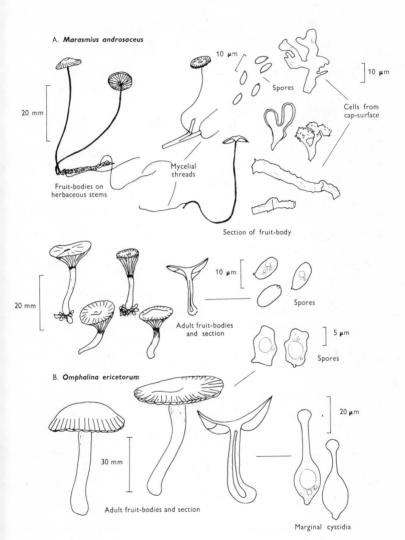

A. *Marasmius androsaceus*

20 mm

10 μm

Spores

10 μm

Cells from
cap-surface

Mycelial
threads

Fruit-bodies on
herbaceous stems

Section of fruit-body

20 mm

10 μm

Spores

Adult fruit-bodies
and section

5 μm

Spores

B. *Omphalina ericetorum*

20 μm

30 mm

Adult fruit-bodies and section

Marginal cystidia

C. *Entoloma helodes*

Gills: white or whitish at first then dirty pinkish brown, adnate and emarginate.

Spore-print: dull salmon-pink.

Spores: medium to long, angular, ellipsoid-oblong, slightly cinnamon-colour under the microscope and $9-12 \times 7-8$ μm in size.

Marginal cystidia: conspicuous, spindle or bottle-shaped and with subcapitate apex.

Facial cystidia: absent.

Hypholoma ericaeum (Fries) Kühner

Cap: width 15–30 mm. *Stem:* width 4–7 mm; length 50–100 mm.

Description:

Cap: fleshy, convex, later becoming flattened but remaining slighty umbonate at the centre, viscid at first, smooth and shining when dry, bright reddish to sand-colour or brown.

Stem: slender, yellow above, brown below, smooth and tough.

Gills: adnate or adnexed, purplish black with a whitish margin and fairly crowded.

Flesh: yellowish or red-brown in the stem.

Spore-print: purple-brown.

Spores: long, dark purple-brown, broadly ellipsoid and $12-15 \times 7-9$ μm in size.

Marginal cystidia: cylindrical or flask-shaped.

Facial cystidia: flask-shaped and filled with contents which become yellowish in solutions containing ammonia.

Clavaria argillacea (Persoon) Fries

Fruit-body: height 20–60 mm.

Description:

Fruit-body: club-shaped, blunt or rounded at the apex, cylindrical or compressed and often grooved, yellow ochraceous or buff.

Stem: distinct but short and yellowish.

Flesh: yellowish.

Spore-print: white.

Spores: medium-sized, hyaline under the microscope, smooth and $10-11 \times 5-6$ μm in size.

All these three species are typical of bare peaty soil, or moss covered peat amongst or around Heather or Ling (*Calluna vulgaris*) bushes.

Plate 78. Moorland, moss-cushion and mountain fungi

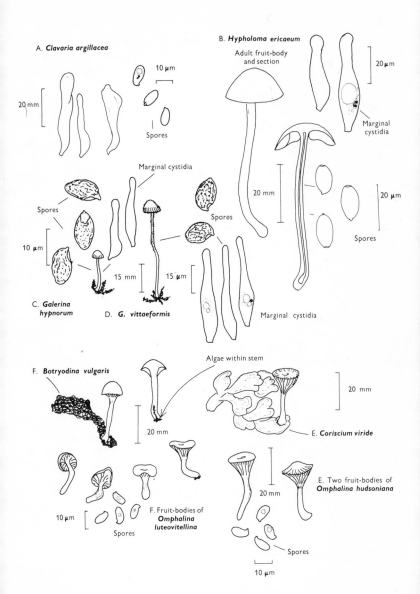

A. *Clavaria argillacea*

10 µm

20 mm

Spores

B. *Hypholoma ericaeum*

Adult fruit-body
and section

20 µm

Marginal
cystidia

20 mm

20 µm

Spores

Marginal cystidia

Spores

10 µm

Spores

C. *Galerina hypnorum*

D. *G. vittaeformis*

15 mm

15 µm

Marginal cystidia

F. *Botryodina vulgaris*

Algae within stem

20 mm

E. *Coriscium viride*

20 mm

E. Two fruit-bodies of
Omphalina hudsoniana

10 µm

Spores

F. Fruit-bodies of
Omphalina luteovitellina

20 mm

10 µm

Spores

Mountain fungi and the so-called Basidiolichens

'Basidiolichens.' Plate 78.

Omphalina ericetorum (Fries) M. Lange has already been described (p. 232): it grows on acidic soils and ascends into mountain areas where it frequently grows on algal scum which accumulates around *Sphagnum* plants.

Under these conditions the algal cells enter the base of the fungus and grow in the cavity of the stem and amongst those hyphae which constitute the base. This association, however, appears to be much closer in the two lichens *Coriscium viride* (Acharius) Vain and *Botrydina vulgaris* Meneghini which have long been classified as species of lichen of unknown affinity because no perfect state was known. *Coriscium viride* consists of blue-green overlapping plates or scales with narrow rounded often paler margins and which dry out greenish brownish grey. *Botrydina vulgaris*, in contrast, consists of dark green, gelatinous blobs drying out greenish brown.

Coriscium is now considered to be an association of an algae and a Basidiomycete, the latter being the agaric, *Omphalina hudsoniana* (Jennings) Bigelow, which resembles *O. ericetorum* but for the pinkish coloured stem. *Botrydina* may be a complex of several separate associations of an algae with different species of *Omphalina*. In the high mountains the association is with *O. luteovitellina* (Pilát & Nannfeldt) M. Lange a small uniformly bright yellow agaric, whilst in *Sphagnum* bogs it is with *O. sphagnicola* (Berkeley) Moser. *Myxomphalia maura* (Fries) Hora, a fungus typical of burnt ground, is also reported to take up this association in lowland woods and *O. velutina* (Quélet) Quélet appears to be capable of forming a loose relationship with algal cells also. This is a most interesting association and research work is still at an early stage. In the tropics and subtropical regions of the world, similar associations are found on rotten and decomposing trunks and stumps. In these examples the *Basidiomycetes* are frequently fairy-clubs, particularly species of *Multiclavula* ('many small clubs'). A few species of this genus may be found also in North temperate woodlands. *Botrydina* also grows in Europe with *Stereum fasciatum* (Schw.) Fries and *Athelia viride* (Bres.) Parm. (see p. 176), and *Odontia bicolor* (Fries) Quélet is rarely collected without green algal cells buried in the thallus. Perhaps associations like this are much commoner than at first supposed.

Probably the most remarkable of this group of poorly known organisms is *Cora pavonia* (Sw.) Fries which produces masses of interlocking fans; it is tropical and found in Brazil.

Mountain fungi: general remarks

There are several groups of mountain fungi, some mycorrhizal formers, some which prefer peaty soil and some which are associated with algae forming a loose relationship—the Basidiolichens. When the mountain top is covered with such dwarf willows as *Salix herbacea* or *S. reticulata* the leaves are cast each year, woody tissue develops above and below the ground; in fact all the processes taking place in our familiar woodlands are also taking place in these communities, the only difference being that the trees are dwarf. Indeed it looks quite odd to see normal sized agarics growing amongst the woody stalks of dwarf trees, the leaves of which are often one-tenth the size of the fruit-bodies, but this is what happens.

The mycorrhizal formers in these conditions include species of *Russula* (e.g. *Russula alpina* Möller & Schaeffer, *R. xerampelina* var. *pascua* Favre (see p. 45), *Lactarius* (e.g. *Lactarius lacunarum* Hora see p. 50), *Cortinarius* (e.g. *C. anomalus* (Fries) Fries see p. 42) and *Amanita* (e.g. *Amanita nivalis* Greville see p. 56). Subterranean fungi are also found, e.g. *Elaphomyces* see p. 244, and, just as woodlands, valley bottoms have a saprophytic ground flora of toadstools so do the high mountain 'woods', and many familiar fungi of the lowerland areas are to be found there also, e.g. *Mycena epipterygia* (Fries) S. F. Gray, *Mycena olivaceo-marginata* (Massee) Massee (see p. 88.)

The barer tops of the mountains, where large areas of moss are only to be found, support species of *Hygrocybe*, e.g. *H. lilacina* (Laestadius) Moser and *H. subviolacea* (Peck) P. D. Orton & Watling (see p. 97).

In the moist atmosphere on the hills in western Scotland, woodland-like floras containing familiar flowering plants are found on the mountain sides often much higher than in central Scotland. It is in such communities that typical woodland fungi are also to be found, e.g. *Nolanea cetrata* (Fries) Kummer (see p. 101).

(vii) Sand-dune fungi

Inocybe dunensis P. D. Orton
Cap: width 27–75 mm. *Stem:* width 4–10 mm; length 35–80 mm.
Description:

Cap: convex then expanded, usually broadly umbonate, pale or dirty ochraceous paler at the margin, reddish brown at the centre, smooth, radially fibrillose towards the margin and sometimes showing the remains of a pale greyish buff veil.

Stem: equal with marginate or rounded bulb at the base, white or whitish, then becoming discoloured pinkish or brownish, powdered with white, at first, but finally silky.

Gills: free or narrowly adnate, subcrowded, whitish then clay-buff, finally snuff-brown with whitish edge.

Flesh: white or whitish, tinted ochraceous or dirty pinkish and with strong smell of rancid oil.

Spore-print: snuff-brown.

Spores: medium to long, ellipsoid-oblong, indistinctly nodulose or wavy-angular and 9–12 × 6–7 μm in size.

Facial cystidia: swollen, spindle-shaped with short, broad neck, thick-walled and crested with crystals.

Marginal cystidia: spindle-shaped and crested with crystals.

General Information: This fungus is often buried to half-way in the sand of slacks near dwarf willows (*Salix* spp.). Three other species of *Inocybe* grow in dune-slacks *I. halophila* Heim, *I. serotina* Peck and *I. devoniensis* P. D. Orton, but all differ in their spores being smooth and elongate-cylindric. *Astrosporina*, a name referring to the shape of the spore, has been considered a genus of agarics in its own right and to this group *I. dunensis* would belong. However, as the members show the same range of characters as those species with the smooth spores it seems unnecessary to split *Inocybe* into two. The cystidia in many species are unusual, being crested with a bundle of crystals which have been reported as being calcium oxalate, although even the simplest school-laboratory tests have been rarely applied to them (see p. 84).

Plate 79. Sand-dune fungi

A. *Inocybe dunensis*

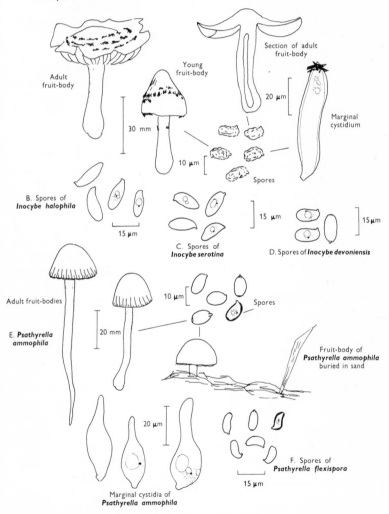

Adult fruit-body

Young fruit-body

Section of adult fruit-body

30 mm

20 μm

Marginal cystidium

10 μm

Spores

B. Spores of *Inocybe halophila*

15 μm

C. Spores of *Inocybe serotina*

15 μm

D. Spores of *Inocybe devoniensis*

15 μm

E. *Psathyrella ammophila*

Adult fruit-bodies

20 mm

10 μm

Spores

Fruit-body of *Psathyrella ammophila* buried in sand

20 μm

Marginal cystidia of *Psathyrella ammophila*

F. Spores of *Psathyrella flexispora*

15 μm

Psathyrella ammophila (Durieu & Léville) P. D. Orton

Sand-dune brittle-cap

Cap: width 20–40 mm. *Stem:* width 4–8 mm; length 40–80 mm.

Description: Plate 79.

Cap: semiglobate to convex, pale dingy clay-colour or dark tan to dirty brownish, non-striate, rather fleshy and usually sand covered.

Stem: deeply rooting in sand and club-shaped towards the base, similarly coloured to the cap except for the whitish apex.

Gills: adnate, subfuscous or dark dirt-brown.

Flesh: dirty buff and with no distinct smell.

Spore-print: pale snuff-brown with purplish flush.

Spores: long, ovoid, yellowish-grey brown under the microscope with a distinct germ-pore and 10–12 × 7 μm in size.

Marginal cystidia: balloon-shaped, obtuse or somewhat bottle-shaped and hyaline.

Facial cystidia: sparse, similar to the marginal cystidia, voluminous.

General Information: This is a very distinct fungus found amongst stems of Marram grass in sand-dune systems. At first sight it appears as if it is growing in the bare sand, but by careful excavation it usually is found attached to pieces of Marram grass, indeed the hyphae enter the roots of the grass, but apparently do not kill them.

This fungus was first described in the genus *Psilocybe* (see p. 114) because of its brownish purple spore-print, but the cap-surface is composed of rounded cells and so is related to all the other species of *Psathyrella*.

Psathyrella flexispora Wallace & P. D. Orton grows in similar habitats amongst *Ammophila* and other seashore grasses. It is easily recognised by the chocolate, umber or date-brown cap and the peculiar shaped spores, which look as if they have been slightly twisted during their development.

Stropharia coronilla (Fries) Quélet, resembling a little mushroom (i.e. *Agaricus*) is also found in sand-dune systems and, just as species of *Psathyrella*, it possesses purplish black spores. However, the cap is ochraceous yellow with a whitish margin formed of veil fragments. The stem is white becoming yellow with age and possesses a narrow, white striate ring. The spores are ellipsoid and measure 8–9 × 4–5 μm and it has filamentous cells in the cap. Unlike *P. ammophila* it is not confined to sand-dune systems but it is also to be found in pastures and on heaths.

Plate 80. Sand-dune fungi

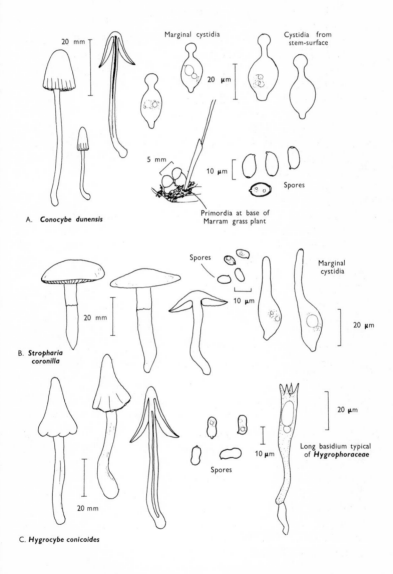

20 mm

Marginal cystidia

Cystidia from stem-surface

20 μm

5 mm

10 μm

Spores

Primordia at base of Marram grass plant

A. **Conocybe dunensis**

Spores

10 μm

Marginal cystidia

20 μm

20 mm

B. **Stropharia coronilla**

20 μm

Long basidium typical of **Hygrophoraceae**

10 μm

Spores

20 mm

C. **Hygrocybe conicoides**

Conocybe dunensis P. D. Orton   Sand-dune brown cone cap.

Cap: width 10–30 mm. *Stem:* width 2–4 mm; length 40–100 mm.

Description: Plate 80.

Cap: conical then conico-expanded, date-brown, dull sand-colour or dark liver-colour, drying buff or ochraceous, expallent, not or indistinctly striate when moist.

Stem: whitish or pale ochraceous then darker ochraceous or dirty brownish from the base up, lower part whitish and buried in the sand.

Flesh: thin and pale ochraceous.

Gills: adnate, whitish but soon pale honey and finally rusty honey.

Spore-print: rust-brown.

Spores: long, ellipsoid or slightly amygdaliform, golden brown under the microscope with large germ-pore and 12–14 × 7–8 μm in size.

Facial cystidia: absent.

Marginal cystidia: capitate.

General Information: C. dunensis differs from C. tenera in its dull colours (see p. 116) and habitat preferences. *Conocybe dunensis*, *Stropharia coronilla*, the two species of *Psathyrella* are all dull-coloured. However, in the sand-dunes colourful agarics are also found. The most common is *Hygrocybe conicoides* (P. D. Orton) Orton & Watling; *Laccaria maritima* (Theodowicz) Moser is indeed an unusual but rewarding find. 'Lac' as in *Laccaria* is a red-brown resinous substrate produced by the lac-insect and resembles the cap colour of many species of the genus, including *L. maritima*, *L. laccata* and *L. proxima* (see p. 86). All these fungi were formerly placed in *Clitocybe*, but they differ in the warted or spiny spores which at maturity give the rather thick gills the appearance of being heavily talced. *L. maritima* can be distinguished from all other species of *Laccaria* by the elongated spores which are minutely spiny and not strongly warted as in *L. laccata*. *Hygrocybe conicoides* (P. D. Orton) Orton & Watling has a conical to conico-convex, acutely umbonate cap with wavy-lobed margin; it is scarlet or cherry-red, discolouring blackish with age or on bruising. The gills are at first chrome-yellow then become flushed red and the stem is yellow or greenish lemon becoming streaky blackish after handling. The spores are 10–13 × 4–5 μm in size and slightly French bean-shaped. It can be readily distinguished from close relatives, e.g. *H. conica* (Fries) Kummer by the gills soon turning reddish, the reddish cap and the narrow spores.

(viii) Subterranean fungi

General notes

The adaptive habit of growing completely submerged beneath the surface of the ground has developed in all the major groups of fungi. Thus the simplest form related to the common bread-mould have taken up the character just as certain relatives of the disc-fungi (discomycetes) and of the flask-fungi (pyrenomycetes). In the higher fungi in several foreign countries even agarics, polypores and stinkhorns have become hypogeous, but in this country we have a very depauparate flora composed of some twenty-eight species of false (Basidiomycete) truffle. The following key may assist in identifying the different groups of hypogeous fungi for some of these species are of commercial value and includes the French or Perigord truffle, *Tuber melanospermum* Vittadini which is used as a constituent of Pâté de Foie Gras, and many of the fungi used as poor quality substitutes. There is a long folk-history surrounding truffles and they have been utilised in the production of aphrodisiacs for centuries. Seeking them out was a difficulty and has been overcome in different countries in different ways. Thus in continental Europe, pigs have been used to sniff them out but on finding them the pigs cannot eat the truffles because of a ring placed through their nose. In Dorset a particular breed of dog was developed to do the same job—the Dorset hounds.

A simple key would read as follows:—

1. Spores produced on basidia 2
   Spores produced in asci 4
2. Chambers throughout the inner tissue containing spores of approximately the same age 3
   Chambers in the inner tissues containing spores found at different stages of development . . . . . . *Hymenogaster*
3. Basidiospores brown or greenish brown under the microscope, and black in mass . . . . *Melanogaster*
   Basidiospores colourless or pale honey colour under the microscope and ochraceous in mass . . . . . . *Rhizopogon*
4. Asci globose, irregularly arranged within the fruit-body and quickly breaking down to shed the spores . . *Elaphomyces*
   Asci globose or club-shaped and arranged in fertile areas which do not rapidly break down to shed the spores . . *Tuber & relatives*

Basidiomycetes

Rhizopogon roseolus (Corda) Fries   Red truffle
Description:

Fruit-body: globular to tubiform and up to 60 mm broad, partly
   covered in mycelial cords, dirty white, later reddish-tawny gradually
   reddish and finally olive-brown, it soon becomes tawny on bruising
   when fresh and young.

Spores: medium sized, narrowly ellipsoid, smooth at first, hyaline
   then pale olive under the microscope and measuring 8–11 × 4 μm.

Habitat & *Distribution:* This fungus is not uncommon on the edges of
   paths, in pine woods just pushing up through the soil surface.

Ascomycetes

Elaphomyces granulatus Fries   Harts' truffle
Description:

Fruit-body: globose to ovoid, 20–40 mm broad, pale ochraceous,
   covered in small pyramidal warts, and when it is cut it shows three
   layers, an outer thin yellowish zone, an inner thicker compact white
   zone and within this a purplish black area full of spores separated
   into chambers by bands of sterile white tissue; the first two zones
   make up the 'rind'.

Spores: spherical, blackish brown, warty, 24–32 μm in diameter; eight
   contained in globose asci.

Habitat & *Distribution:* This fungus is not uncommon in the surface
   layers of pine woods at the junction of needle debris and mineral
   soil. *E. muricatus* Fries is similar, but differs in the marbled flecked
   interior.

Tuber aestivum Vittadini   English truffle
Description:

Fruit-body: subglobose except for basal flattening, up to 80 mm
   broad, covered in 5–6-sided pyramidal scales, dark brown to violaceous, white then greyish brown within, separated by a network of
   veins radiating from the basal cavity.

Spores: very large, ellipsoid, light or yellowish brown and ornamented
   with a prominent network, borne in two's and sixes in subglobose
   asci and variable in size, 20–40 × 15–30 μm.

A. *Rhizopogon roseolus*

30 mm

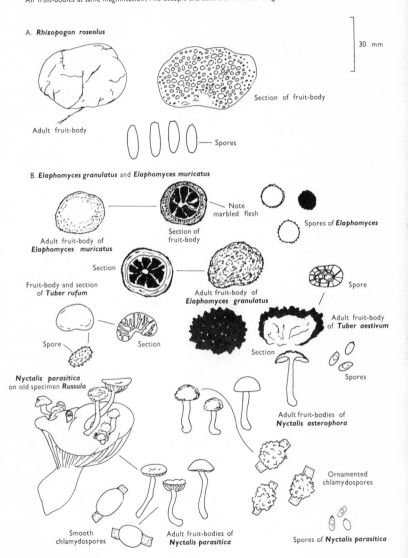

Section of fruit-body

Adult fruit-body

Spores

B. *Elaphomyces granulatus* and *Elaphomyces muricatus*

Note marbled flesh

Spores of *Elaphomyces*

Adult fruit-body of *Elaphomyces muricatus*

Section of fruit-body

Section

Fruit-body and section of *Tuber rufum*

Adult fruit-body of *Elaphomyces granulatus*

Spore

Spore

Section

Adult fruit-body of *Tuber aestivum*

Section

Spores

Nyctalis parasitica on old specimen *Russula*

Adult fruit-bodies of *Nyctalis asterophora*

Ornamented chlamydospores

Smooth chlamydospores

Adult fruit-bodies of *Nyctalis parasitica*

Spores of *Nyctalis parasitica*

Habitat & Distribution: This fungus is to be found buried in the surface layers of soil in beech woods. *T. rufum* is smaller and smoother and the spores are not crested but simply minutely spiny.

Illustrations: R. luteolus—Hvass 322; LH 215. *El. granulatus*—Hvass 325; LH 49. *T. aestivum*—LH 43. *Melanogaster variegatus*—LH 215 (see p. 243). *Hymenogaster tener*—LH 215 (see p. 243).

(ix) Fungal parasites

Nyctalis parasitica (Fries) Fries   Pick-a-back-toadstool
Description: Plate 81.

Cap: bell-shaped then becoming expanded, silky dirty white, but gradually grey with a flush of lilac with age.

Stem: slender, white and smooth except for the base.

Gills: pallid but soon becoming brownish, adnate or adnate with tooth, thick and distant alternately long and short and contorted or united with age.

Flesh: dark brown.

Spore-print: buff.

Spores: small, hyaline under the microscope, ovoid, 5–6 × 3–4 μm but usually replaced completely or in part by ovoid, smooth, thick-walled and pale brownish asexually produced spores (chlamydospores) measuring about 15 × 10 μm in size.

Habitat & Distribution: This fungus grows in clusters on old decaying specimens of various species of *Russula* and *Lactarius* (Russulaceae)—see p. 45.

General Information: N. asterophora Fries is closely related and also grows on decaying specimens of various species of *Russula*, particularly *R. nigricans* (Fries) Fries. It differs, however, in the cap being fawn-coloured and very mealy when touched; it is recognised by the poorly formed often developmentally hindered gills on which chlamydospores are formed. Unlike the smooth asexual spores in *N. parasitica* this species has chlamydospores with conical, blunt humps—i.e. star-shaped; *asterophora* in fact means 'I bear stars'. These fungi have been associated by some mycologists with the common chanterelle (*Cantharellus cibarius* Fries, see p. 162) in virtue of them possessing reduced fold-like gills. However, the fold-like gills are secondary in nature, correlated with the active production of chlamydospores and the supression of the formation of basidio-spores. The gills are not therefore of a primitive type. The genus *Nyctalis* is related to fungi such as *Tephrocybe palustris* (Peck) Donk (see p. 223).

There are several rather uncommon 'agaric-parasites' of agarics or other higher fungi, e.g. *Volvariella surrecta* (Knapp) Singer, but their formal description must be left to other more advanced texts. However, the intriguing bolete, *Boletus parasiticus* Fries, which grows on *Scleroderma* (earth-balls) in this country has been mentioned and figured previously (p. 35 & Plate 64). It is of interest to note that a close relative of *B. parasiticus* in Japan lives on another group of Gasteromycetes.

Illustrations: N. parasitica—F 11[a]; LH 81; WD 25[7]. *N. asterophora*—LH 81; WD 25[8].

General notes on Fungicoles

Many beginners are confused on finding specimens which, although appearing agaric-like, are covered in long hairs or irregularly shaped bumps. Indeed many of these abnormalities are true agarics attacked by microscopic fungi, and I know of one textbook on mushrooms and toadstools which includes such an abnormality amongst the discussion on the normal fruit-bodies. Thus *Sporodinia grandis* Link, which is a primitive fungus, attacks many fungi reducing them to a grey velvety mass of fungal filaments. Specimens of several species of *Mycena* (p. 88) are common in autumn, covered in whiskers with small nobbles on the top. These whiskers are produced by the parasitic *Spinellus megalocarpus* (Corda) Karsten, another primitive fungus—a phyco-mycete.

In some wet seasons the orange and green coloured *Lactarius deliciosus* (Fries) S. F. Gray is to be found contorted and covered in small pinkish to lilac pimples of the ascomycete *Byssonectria lactaria* (Fries) Petch, and other species of *Lactarius* are attacked by *Byssonectria viridis* (Berkeley & Broome) Petch which converts the fruit-bodies into a hardened mass of green tissue. In North America, species of *Lactarius* are frequently attacked by *Hypomyces lactifluorum* (Schweintz) Tulasne and the whole fungus is reduced to a contorted acidic-smelling mass of fungal tissue with vivid orange pimples or warts on the outer surface. These parasitic fruit-bodies are eaten as a delicacy in their own right whereas the same consumer will be less enthusiastic about eating the same agaric before it is so deformed.

Boletes particularly *B. subtomentosus* Fries, *B. chrysenteron* St Amans and *B. edulis* Fries are frequently converted into yellow powdery masses due to the production of asexual spores of the fungus *Sepedonium chrysospermum* Fries; the sexual stage occurs on the remains after they have collapsed into the soil surface—this stage is called *Apiocrea chrysosperma* (Tulasne) Sydow. Several closely related fungi in the genus *Hypomyces* also attack agarics.

The yellow pustules found on the spore-bearing surface of the birch polypore *Piptoporus* (p. 142) is *Hypocrea pulvinata* Fuckel; it is only one of several lower fungi which grow on bracket fungi. The genus *Cordyceps* has been mentioned previously (p. 206) and in the discussion it was indicated that certain hypogeous fungi are attacked by members of this genus.

White gelatinous pustules found amongst the fruit-bodies of *Stereum sanguinolentum* (p. 176) have a hard white centre. On examination these 'nuclei' are aborted structures of the stereum covered in the jelly-fungus *Tremella encephala* Persoon. This fungus is apparently parasitic; it is closely related to *Tremella foliacea* and *T. mesenterica* described on page 184.

G. APPENDIX

(i) Species list of specialised habitats

INTRODUCTION

Although some fungi prefer one type of woodland more than another many fungi are less specialised and may be found in all kinds of woods. Indeed many fungi which we usually associate with a woodland fungus flora can also be commonly seen in pastures and gardens, e.g. *Laccaria laccata* (Fries) Cooke, *Hygrophoropsis aurantiaca* (Fries) Maire.

It is useful to consider the fungi of different woodland types separately, but this in some cases is very difficult because some species are not exclusive; indeed some species may grow in completely contrasting habitats, e.g. *Amanita muscaria* (Fries) Hooker in both birch and conifer woods, or on contrasting substrates, e.g. *Fomes fomentarius* (Fries) Kickx on birch in Scotland and beech on the continent of Europe. The picture becomes even more complex because frequently woods, in fact, often include several tree species growing in close proximity and it is then difficult to draw connections between a fungus and the tree with which it is truly growing—we know little or nothing except for mycorrhizal fungi, why certain fungi prefer certain habitats.

A parallel example is that phenomenon seen in certain polypores which only attack twigs or branches and not stumps or trunks, whilst others grow exclusively on stumps. We know little of the reasons for these demarcations, even when they occur within the same host. Mycology, therefore, offers to the beginner and the professional many opportunities in physiology and ecology.

In grassland areas it is difficult to know where to draw the line between one plant-community and another when listing species, for although ecologically distinct both would come under the name grassland. In the field, however, this is often very obvious and there is little doubt that fungi can give just as accurate an indication as to the soil-type, as many mosses or vascular plants. In sand-dune systems, the mobile dunes offer a different ecological niche to that of the fixed dunes which in many ways resemble grasslands. Thus although the lists below are split into easily manageable units, some flexibility must still be allowed. It is meant only as a guide—and will differ in some cases from one place to another, even within the British Isles.

General Woodland

Agaricus silvicola (Vitt.) Peck
Amanita citrina S. F. Gray
A. excelsa (Fries) Kummer
A. rubescens (Fries) S. F. Gray
A. vaginata (Fries) Vittadini
Boletus calopus Fries
B. erythropus (Fries) Secretan
B. piperatus Fries
Cantharellus infundibuliformis Fries
Clitocybe clavipes (Fries) Kummer
C. fragrans (Fries) Kummer
C. nebularis (Fries) Kummer
C. odora (Fries) Kummer
Collybia butyracea (Fries) Kummer
C. confluens (Fries) Kummer
C. dryophila (Fries) Kummer
Hebeloma crustuliniforme (St Amans) Quélet
Hygrocybe strangulata (Orton) Moser
Hygrophoropsis aurantiaca (Fries) Maire
Inocybe eutheles (Berkeley & Broome) Quélet
I. fastigiata (Fries) Quélet
I. geophylla (Fries) Kummer
Laccaria laccata (Fries) Cooke
Lactarius mitissimus (Fries) Fries
L. piperatus (Fries) S. F. Gray
L. subdulcis (Fries) S. F. Gray
Limacella glioderma (Fries) Maire
Mycena filopes (Fries) Kummer
M. galopus (Fries) Kummer
M. pura (Fries) Kummer
M. sanguinolenta (Fries) Kummer
M. vitilis (Fries) Quélet
Paxillus involutus (Fries) Fries
Ripartites tricholoma (Fries) Karsten
Russula adusta (Fries) Fries
R. atropurpurea (Krombholz) Britz.
R. delica Fries
R. foetens (Fries) Fries
R. nigricans (Mérat) Fries

R. ochroleuca (Secretan) Fries
R. xerampelina (Secretan) Fries
Tricholoma argyraceum (St Amans) Gillet
T. orirubens Quélet
T. saponaceum (Fries) Kummer
T. sciodes (Secretan) Martin
T. terreum (Fries) Kummer
T. virgatum (Fries) Kummer
Tylopilus felleus (Fries) Karsten

Hydnum repandum Fries

Phallus impudicus Persoon
Scleroderma citrinum Persoon
S. verrucosum Persoon

Leotia lubrica Persoon
Microglossum viride (Fries) Gillet

On wood

Armillaria mellea (Fries) Kummer
Crepidotus variabilis (Fries) Kummer
Hypholoma fasciculare (Fries) Kummer
H. sublateritium (Fries) Quélet
Pluteus cervinus (Fries) Kummer

Calocera cornea (Fries) Fries
Coriolus versicolor (Fries) Quélet
Merulius tremellosus Fries
Schizophyllum commune Fries
Stereum hirsutum (Fries) Fries
S. rugosum (Fries) Fries

Lycoperdon pyriforme Persoon

Coryne sarcoides (S. F. Gray) Tulasne
Cudoniella acicularis (Fries) Schroeter
Nectria cinnabarina (Fries) Fries
Xylosphaera hypoxylon Dumortier
X. polymorpha (Mérat) Dumortier

Conifer Woods

characterised by species of *Suillus*, *Chroogomphus*, *Gomphidius*, several *Lactarius* and *Russula* spp.

Agaricus sylvatica Secretan
Amanita porphyria (Fries) Secretan
Boletus badius Fries
B. pinicola Venturi
Chroogomphus rutilus (Fries) O. K. Miller
Clitocybe flaccida (Fries) Kummer
C. langei Hora
Collybia distorta (Fries) Quélet
Cortinarius callisteus (Fries) Fries
C. gentilis (Fries) Fries
C. mucosus (Fries) Kickx
C. pinicola P. D. Orton
C. sanguineus (Fries) Fries
C. semisanguineus (Fries) Gillet
Cystoderma amianthinum (Fries) Fayod
Gomphidius glutinosus (Fries) Fries
G. maculatus Fries
G. roseus (Fries) Karsten
Hygrophorus hypothejus (Fries) Fries
Hypholoma marginatum (Fries) Schroeter
Inocybe calamistrata (Fries) Gillet
Lactarius camphoratus (Fries) Fries
L. deliciosus (Fries) S. F. Gray
L. helvus (Fries) Fries
L. rufus (Fries) Fries
Leccinum vulpinum Watling
Marasmius androsaceus (Fries) Fries
Mycena adonis (Fries) S. F. Gray (= *Hemimycena*)
M. amicta (Fries) Quélet
M. capillaripes Peck
M. coccinea Quélet
M. rubromarginata (Fries) Kummer
M. vulgaris (Fries) Kummer
Nolanea cetrata (Fries) Kummer
N. cuneata Bresadola
Rozites caperata (Fries) Karsten
Russula caerulea Fries
R. decolorans (Fries) Fries
R. emetica (Fries) S. F. Gray
R. erythropus Peltereau

R. nauseosa (Secretan) Fries
R. obscura Romell
R. paludosa Britz.
R. queletii Fries
R. sardonia Fries
Tricholoma albobrunneum
T. flavovirens (Fries) Lundell
T. focale (Fries) Ricken
T. imbricatum (Fries) Kummer
T. vaccinum (Fries) Kummer

Ramaria ochraceo-virens (Jungh.) Donk
R. invallii (Cotton & Wakef.) Donk
Sarcodon imbricatum (Fries) Karsten
Sparassis crispa (Wulfen) Fries
Thelephora palmata (Bulliard) Patouillard
T. terrestris Fries

Geastrum pectinatum Persoon

Hypogeous

Rhizopogon luteolus Fries
Elaphomyces granulatus Fries
E. muricatus Fries

On cones

Baeospora myosura (Fries) S nger
Strobilurus esculentus (Wulf. ex Fr.) Singer
S. stephanocystis (Hora) Singer
S. tenacellus (Fries) Singer

Auriscalpium vulgare S. F. Gray

On conifer wood

Gymnopilus penetrans (Fries) Murrill
Hypholoma capnoides (Fries) Kummer
Mycena alcalina (Fries) Kummer
Lentinus tigrinus (Fries) Fries
Paxillus atrotomentosus (Fries) Fries

P. panuoides (Fries) Fries
Pholiota flammans (Fries) Kummer
Pleurotellus porrigens (Fries) Singer
(= *Pleurocybella*)
Pluteus atromarginatus Kühner
Tricholompsis rutilans (Fries) Singer
Xeromphalina campanella (Fries) Maire

Calocera viscosa (Fries) Fries
Dacrymyces stillatus Nees ex Fries
Pseudohydnum gelatinosum (Fries) Karsten
Gloeophyllum sepiarium (Fries) Karsten
Heterobasidion annosum (Fries) Brefeld
Hirschioporus abietinus (Fries) Donk
Laetiporus sulphureus (Fries) Murrill
Phaeolus schweinitzii (Fries) Patouillard
Stereum sanguinolentum (Fries) Fries
Tremella encephala Persoon
T. foliacea (Persoon) Persoon
Tyromyces stipticus (Fries) Kotlaba & Pouzar

Deciduous Woods General

Amanita fulva Secretan
A. inaurata Secretan
A. virosa Secretan
Boletus edulis Fries
B. chrysenteron St Amans
B. luridus Fries
B. subtomentosus Fries
Collybia peronata (Fries) Kummer
Lactarius vellereus (Fries) Fries
Russula cyanoxantha (Secretan) Fries
R. grisea (Secretan) Fries
R. heterophylla (Fries) Fries
R. lutea (Fries) Fries
R. ochroleuca (Secretan) Fries
Tricholoma album (Fries) Kummer
T. columbetta (Fries) Kummer
T. saponaceum (Fries) Kummer
T. sulphureum (Fries) Kummer

Cantharellus cibarius Fries

Clavulina cinerea (Fries) Schroeter
C. cristata (Fries) Schroeter
Hydnum repandum Fries

Geastrum rufescens Persoon
Lycoperdon perlatum Persoon

Helvella crispa Fries
H. elastica (St Amans) Boudier
H. lacunosa Fries
Disciotis venosa (Persoon) Boudier
Paxina acetabulum (St Amans) Kuntze
Peziza badia Mérat
P. succosa Berkeley

On wood

Coprinus disseminatus (Fries) S. F. Gray
C. micaceus (Fries) Fries
Crepidotus mollis (Fries) Kummer
Galerina mutabilis (Fries) P. D. Orton
Gymnopilus junonius (Fries) P. D. Orton
Mycena galericulata (Fries) S. F. Gray
Oudemansiella radicata (Fries) Singer
Pholiota squarrosa (Fries) Kummer
Pleurotoid fungi (see p. 74)
Psathyrella candolleana (Fries) R. Maire
P. hydrophilum (Mérat) Maire

Coniophora puteana (Fries) Karsten
Meripilus giganteus (Fries) Karsten
Tremella mesenterica Hooker

Beech Woods

Amanita citrina var alba Gillet
Boletus edulis Fries
B. satanus Lenz
Collybia fuscopurpurea (Fries) Kummer
Coprinus picaceus (Fries) S. F. Gray
Cortinarius pseudosalor J. Lange
C. bolaris (Fries) Fries

Hygrophorus chrysaspis Métrod
Laccaria amethystea (Mérat) Murrill
Lactarius blennius (Fries) Fries
L. pallidus (Fries) Fries
L. tabidus Fries
Marasmius cohaerens (Fries) Cooke &
  Quélet
M. wynnei Berkeley & Broome
Mycena capillaris (Fries) Kummer
  (on leaves)
M. pelianthina (Fries) Quélet
Russula alutacea (Fries) Fries
R. fellea (Fries) Fries
R. lepida Fries
R. mairei Singer
R. virescens (Zantedschi) Fries
Tricholoma ustale (Fries) Kummer

Clavariadelphus   pistillaris   (Fries)
  Donk
Geaster triplex Jungh
G. fimbriatum Fries

Hypogeous

Melanogaster variegatus Vittadini
Tuber aestivum Vittadini

On wood

Oudemansiella mucida (Fries) Höhnel
O. radicata (Fries) Singer
Panus torulosus (Fries) Fries
Pholiota adiposa (Fries) Kummer
Stropharia squamosa (Fries) Quélet

Bjerkandera adusta (Fries) Karsten
Datronia mollis (Fries) Donk
Hiericium coralloides (Fries) S. F.
  Gray
Lentinellus cochleatus (Fries) Karsten
Pseudotrametes gibbosa (Fries) Bond.
  & Singer

Bulgaria inquinans Fries (a large dark
  brown, gelatinous discomycete)
Several pyrenomycetes are recorded
  and dealt with by J. Webster in a

popular account published in *The
Naturalist*, London 1953, pp. 1–16.

Birch Woods

Amanita crocea (Quélet) Kühner &
  Romagnesi
Boletus edulis Fries
Cortinarius armillatus (Fries) Fries
C. crocolitus Quélet
C. hemitrichus (Fries) Fries
Lactarius glaucescens Crossland
L. glyciosmus (Fries) Fries
L. lacunarum Hora
L. torminosus (Fries) S. F. Gray
L. turpis (Weinm.) Fries
L. uvidus (Fries) Fries
L. vietus (Fries) Fries
Leccinum holopus (Rostkovius) Watling
L. roseofractum Watling
L. scabrum (Fries) S. F. Gray
L. variicolor Watling
L. versipellis (Fries & Hök) Snell
Russula aeruginea Lindblad ex Fries
R. betularum Hora
R. claroflava Grove
R. gracillima J. Schaeffer
R. nitida (Fries) Fries
R. pulchella Borszczow
R. versicolor J. Schaeffer
Tricholoma fulvum (Fries) Saccardo

On wood

Fomes fomentarius (Fries) Kickx
Lenzites betulina (Fries) Fries
Piptoporus betulinus (Fries) Karsten

Oak Woods

Amanita phalloides (Fries) Secretan
Boletus albidus Rocques
B. appendiculatus Fries
B. pulverulentus Opatowski
B. reticulatus Boudier
B. versicolor Rostkovius
Gyroporus castaneus (Fries) Quélet
Hygrophorus eburneus (Fries) Fries

Lactarius chrysorheus Fries
L. quietus (Fries) Fries
Leccinum quercinum (Pilát) Green &
    Watling
Russula vesca Fries
Tricholoma acerbum (Fries) Quélet

Hypogeous

Hymenogaster   tener   Berkeley   &
    Broome

On wood

Mycena inclinata (Fries) Quélet
Psathyrella obtusata (Fries) A. H.
    Smith

Daedalea quercina Persoon
Fistulina hepatica Fries
Hymenochaete   rubiginosa   (Fries)
    Léville
Peniophora quercina (Fries) Cooke
Inonotus dryadeus (Fries) Murrill
Stereum gausapatum (Fries) Fries

Specific Tree Species

Alder

Lactarius obscuratus (Lasch) Fries
Naucoria escharoides (Fries) Kummer
N. scolecina (Fries) Quélet

On wood

Clavariadelphus fistulosus var. *contorta*
    (Fries) Corner
Exidia glandulosa (St Amans) Fries
Inonotus radiatus (Fries) Karsten
Plicaturiopsis crispa (Fries) Reid

Ash
On wood

Inonotus hispidus (Fries) Karsten
Daldinia concentrica (Fries) Cesati &
    de Notaris

Elder
On wood

Hirneola auricula-judae (St Amans)
    Berkeley
Hyphodontia sambuci (Fries) J. Eriks-
    son

Elm
On wood

Lyophyllum ulmarius (Fries) Kühner
Rhodotus palmatus (Fries) Maire
Volvariella bombycina (Fries) Singer
Rigidoporus ulmarius (Fries) Imaz

Hazel

Lactarius pyrogalus (Fries) Fries
Leccinum carpini (R. Schulzer) Reid

On wood

Hymenochaete corrugata (Fries) Léville
Sarcoscypha coccinea (Fries) Lambotte
    (red   discomycete   occurring   in
    early spring)

Hawthorn

Entoloma clypeatum (Fries) Kummer

On wood

Pholiota squarrosa (Fries) Kummer
Phellinus pomaceus (Persoon) Maire
Stereum purpureum (Fries) Fries

Hornbeam

Lactarius circellatus Fries
Leccinum carpini (R. Schulzer) Reid

Poplar

Lactarius controversus (Fries) Fries
Leccinum aurantiacum (Fries) S. F.
    Gray

L. duriusculum (Schulzer) Singer
Mitromorpha hybrida (Fries) Léville

On wood

Agrocybe cylindracea (Fries) Maire
Pholiota destruens (Brondeau) Gillet
Bjerkandera fumosa (Fries) Karsten
Oxyporus populinus (Fries) Donk

Willow

Hebeloma leucosarx P. D. Orton
H. mesophaeum (Persoon) Quélet
H. testaceum (Fries) Quélet
Lactarius lacunarum Hora

On wood

Daedaleopsis rubescens (Fries) Schroeter
Pluteus salicinus (Fries) Kummer
Phellinus igniarius (Fries) Quélet
Trametes suaveolens (Fries) Fries

Grasslands

Agaricus arvensis Secretan
A. campestris Fries
A. macrosporus (Moëller & Schaeffer) Pilát
Agrocybe semiorbicularis (St Amans) Fayod
Calocybe gambosum (Fries) Singer
C. carneum (Fries) Kummer
Cantharellula umbonata (Fries) Singer
Clitocybe dealbata (Fries) Kummer
C. ericetorum Quélet
C. rivulosa (Fries) Kummer
Clitopilus prunulus (Fries) Kummer
Dermoloma atrocinereum (Fries) P. D. Orton
D. cuneifolium (Fries) Singer
Entoloma porphyrophaeum (Fries) Karsten
Hygrocybe aurantiosplendens R. Haller
H. berkeleyi (P. D. Orton) Orton & Watling
H. chlorophana (Fries) Karsten

H. coccinea (Fries) Kummer
H. conica (Fries) Kummer
H. calyptraeformis (Berkeley & Broome) Fayod
H. flavescens (Kauffman) Singer
H. marchii (Bresadola) Singer
H. nivea (Fries) Orton & Watling
H. nitrata (Pers.) Wunsche
H. obrussea (Fries) Fries
H. pratensis (Fries) Donk
H. psittacina (Fries) Wunsche
H. punicea (Fries) Kummer
H. reai (Maire) J. Lange
H. russocoriacea (Berkeley & Miller) Orton & Watling
H. splendidissima (P. D. Orton) Moser
H. unguinosa (Fries) Karsten
H. virginea (Fries) Orton & Watling
Lepiota procera (Fries) S. F. Gray
Lepista luscina (Fries) Singer
L. saeva (Fries) P. D. Orton
Leptonia griseocyanea (Fries) P. D. Orton
L. incana (Fries) Gillet
L. sericella (Fries) Barbier
L. serrulata (Fries) Kummer
Leucoagaricus naucina (Fries) Singer
Melanoleuca strictipes (Karsten) J. Schaeffer
Mycena flavoalba (Fries) Quélet
M. leptocephala (Fries) Gillet
M. fibula (Fries) Kühner
M. swartzii (Fries) A. H. Smith
Nolanea papillata Bresadola
N. sericea (Mérat) P. D. Orton
N. staurospora Bresadola
Psathyrella atomata (Fries) Quélet
Rhodocybe popinalis (Fries) Singer

Clavaria fumosa Fries
C. vermicularis Fries
Clavulinopsis corniculata (Fries) Corner
C. fusiformis (Fries) Corner
C. helvola (Fries) Corner

Bovista nigrescens Persoon
B. plumbea Persoon

Calvatia utriformis (Fries) Jaap
C. excipuliformis (Fries) Perdeck
Corynetes atropurpureus (Fries) Durand
Geoglossum cookeianum Nannfeldt
G. glutinosum Fries
G. nigritun Cooke
Trichoglossum hirsutum (Fries) Boudier

Lawns: Wasteland: Hedgerows

Agaricus hortensis (Cooke) Pilát
A. bisporus (J. Lange) Pilát
A. xanthodermus Genevier
Agrocybe dura (Fries) Singer
A. erebia (Fries) Kühner
A. praecox (Fries) Fayod
Coprinus comatus (Fries) S. F. Gray
C. acuminatus (Romagnesi) P. D. Orton
C. atramentarius (Fries) Fries
C. micaceus (Fries) Fries
C. plicatilis (Fries) Fries
Flammulaster granulosa (J. Lange) Watling
Lacrymaria velutina (Fries) Konrad & Maublanc
Lepiota cristata (Fries) Kummer
L. friesii (Lasch) Quélet
L. rhacodes (Vittadini) Quélet
Lepista nuda (Fries) Cooke
L. sordida (Fries) Singer
Lyophyllum connatum (Fries) Singer
L. decastes (Fries) Singer
Marasmius oreades (Fries) Fries
Melanophyllum echinatum (Fries) Singer
Mycena olivaceomarginata (Massee) Massee
M. fibula (Fries) Kühner
M. swartzii (Fries) A. H. Smith
Panaeolus fimicola (Fries) Quélet
P. foenisecii (Fries) Schroeter
Psathyrella gracilis (Fries) Quélet
P. squamosa (Karsten) Moser
Tubaria furfuracea (Fries) Gillet
T. pellucida (Fries) Gillet
Volvariella speciosa (Fries) Singer

Langermannia gigantea (Persoon) Lloyd

Aleuria aurantia (Fries) Fuckel
Morchella esculenta St Amans
Verpa conica Persoon

On herbaceous material

Coprinus urticicola (Berkeley & Broome) Buller
Panaeolus subbalteatus (Berkeley & Broome) Saccardo (in middens)

Crucibulum laeve (de Candolle) Kambly
Cyathus olla Persoon

Helicobasidium brebissonii (Desmazieres) Donk

Pistillaria micans (Persoon) Fries
P. quisquilliaris Fries (on bracken stems)

In greenhouses

Lepiota rhacodes var. *hortensis* Pilát
Leucocoprinus cepaestipes (Fries) Patouillard
L. birnbaummii (Corda) Singer
L. brebissonii (Godey) Locquin
L. denudatus (Rabenhorst) Singer
L. lilacinogranulosus (Henning) Locquin
Psilocybe cyanescens Wakefield

Near out-buildings, stables, etc.

Anthurus archeri (Berkeley) E. Fischer
Asteroe ruber La Billardiere
Clathrus ruber Persoon
Lysurus australiensis Cooke & Massee
Queletia mirabilis Fries

Specialised habitats

(a) Dung

Bolbitius vitellinus (Fries) Fries
Conocybe coprophila (Kühner) Kühner
C. pubescens (Gillet) Kühner
C. rickenii (J. Schaeffer) Kühner
Coprinus cinereus (Fries) S. F. Gray
C. ephemeroides (Fries) Fries
C. macrocephalus (Berkeley) Berkeley
C. patouillardii Quélet
C. narcoticus (Fries) Fries
C. niveus (Fries) Fries
C. pellucidus Karsten
C. pseudoradiatus Kühner & Josserand
C. radiatus (Fries) S. F. Gray
Panaeolus semiovatus (Fries) Lundell
P. sphinctrinus (Fries) Quélet
Psathyrella coprobia (J. Lange) A. H. Smith
Psilocybe coprophila (Fries) Kummer
P. merdaria (Fries) Quélet
Stropharia semiglobata (Fries) Quélet

Pyrenomycetes: Genera—*Sordaria; Podospora; Sporormia; Delitschia.*
Discomycetes: Genera—*Cheilymenia; Ascobolus; Coprobia.*
A key to the common dung fungi is given in *Bull. British Myc. Society*, 1968 by Watling & Richardson.

(b) Burnt patches

Aureoboletus cramesinus (Secretan) Watling
Coprinus angulatus Peck
C. lipophilus Romagnesi & Heim
Hebeoloma anthracophilum Maire
Mycena leucogala (Cooke) Saccardo
Myxomphalina maura (Fries) Hora
Pholiota highlandensis (Peck) A. H. Smith
Psathyrella pennata (Fries) Pearson & Dennis
Tephrocybe arthracophila (Lasch) P. D. Orton

T. ambusta (Fries) Donk
T. atrata (Fries) Donk

Coltricia perennis (Fries) Murrill

Anthracobia macrocystis (Cooke) Boudier
A. maurilabra (Cooke) Boudier
A. melaloma (Fries) Boudier
Ascobolus carbonarius Karsten
Geopyxis carbonaria (Fries) Saccardo
Lamprospora astroidea (Hazslinzky) Boudier
Peziza echinospora Karsten
P. petersii Berkeley & Curtis
P. praetervisa Bresadola
P. violacea Persoon
Pyronema omphalodes (St Amans) Fuckel
Tricharia gilva Boudier
Trichophaea woolhopeia (Cooke & Phillips) Boudier

(c) Sand-dunes

Agaricus bernardii Quélet
A. devoniensis P. D. Orton
Conocybe dunensis P. D. Orton
Eccilia nigella Quélet
Hygrocybe conicoides P. D. Orton
Inocybe devoniensis P. D. Orton
I. dulcamara (Persoon) Kummer
I. dunensis P. D. Orton
I. halophila Heim
I. serotina Peck
Laccaria maritima (Theodowicz) Singer
Psathyrella ammophila (Durieu & Léville) P. D. Orton
Stropharia albocyanea (Desmariezes) Quélet

Geaster striatum de Candolle
Tulostoma brumale Persoon
Vascellum depressum (Bonorden) Smarda
Phallus hadriani Persoon

Corynetes arenarius (Rostrup) Durand
Peziza ammophila Durieu & Montagne

(d) Heathland

Cystoderma amianthinum (Fries)
Fayod
Entoloma helodes (Fries) Kummer
E. madidum (Fries) Gillet
Galerina mniophila (Lasch) Kühner
G. praticola (Moëller) P. D. Orton
G. vittaeformis (Fries) Moser
Hygrophoropsis aurantiaca (Fries)
Maire
Hygrocybe cantharella (Schweintz)
Murrill
H. lacma (Fries) Orton & Watling
H. laeta (Fries) Kummer
H. ovina (Fries) Kühner
H. subradiata (Secretan) Orton &
Watling
H. turunda (Fries) Karsten
Hypholoma ericaeum (Fries) Kühner
H. subericaeum (Fries) Kühner
Mycena epipterygia (Fries) S. F.
Gray
M. olivaceomarginata (Massee) Massee
Omphalina velutina (Quélet) Quélet

Clavaria argillacea (Persoon) Fries

Lycoperdon foetidum Bonorden

(e) Marshes

Cortinarius uliginosus Berkeley
Coprinus friesii Quélet (on grass-stems)
C. martinii P. D. Orton (on *Juncus*)
Entoloma sericatum (Britz.) Saccardo
(under birches)
Galerina jaapii Smith & Singer
G. paludosa (Fries) Kühner
G. sphagnorum (Fries) Kühner
G. tibiicystis (Atkinson) Kühner
Hygrocybe cantharella (Schweinitz)
Murrill

H. coccineocrenata (P. D. Orton)
Moser
H. turunda (Fries) Karsten
Hypholoma elongatum (Fries) Ricken
H. udum (Fries) Kühner
Laccaria proxima (Boudier) Patouillard
Marasmius menieri Boudier on *Typha*
Mycena belliae (Johnston) P. D.
Orton on *Phragmites*
M. bulbosa (Cejp) Kühner on *Juncus*
M. integrella (Fries) S. F. Gray on
Cladium
Omphalina ericetorum (Fries) M.
Lange
O. oniscus (Fries) Quélet
O. philonotis (Lasch) Quélet
O. sphagnicola (Berkeley) Moser
Pholiota myosotis (Fries) Singer
Psathyrella sphagnicola (Maire) Favre
Tephrocybe palustris (Peck) Donk

Cudoniella clavus (Fries) Dennis
Mitrula paludosa Fries
Scutellinia scutellata (St Amans) Lambotte (with bright red disc and
conspicuous brown hairs at the
margin)
Vibrissea truncorum Fries (an orange-capped fungus with a black stem)

(f) Mountain tops

Amanita nivalis Greville
Cortinarius anomalus (Fries) Fries
C. cinnamomeus (Fries) Fries
C. tabularis (Fries) Fries
Russula alpina (Blytt) Moëller &
Schaeffer
R. xerampelina var. *pascua* Favre

(g) Mossy areas on the ground, rocks
or stumps

Galerina hypnorum (Fries) Kühner
G. mniophila (Lasch) Kühner
G. mycenopsis (Fries) Kühner
G. praticola Moëller

G. unicolor (Sommerf.) Singer (often on wood)

Leptoglossum lobatus (Fries) Ricken

L. retirugis (Fries) Kühner & Romagnesi

Mycena corticola (Fries) Ricken (on wood)

M. hiemalis (Fries) Quélet (on wood)

M. olida Bresadola (on wood)

Omphalina rickenii Hora

Cyphella muscigena (Pers.) Fries

Cyphellostereum levis (Fries) Reid

Neottiella rutilans (Fries) Dennis

(h) Hypogeous fungi

Melanogaster variegatus Vittadini

Rhizopogon luteolus Fries

R. rubescens Tulasne

Elaphomyces granulatus Fries

E. muricatus Fries

Gyrocratera ploettneriana Hennings

Hydnotrya tulasnei Berkeley & Broome

Melanogaster variegatus Vittadini

Tuber aestivum Vittadini

T. rufum Fries

(i) On rotten fungi

Nyctalis asterophora Fries

N. parasitica (Fries) Fries

Collybia cirrhata (Fries) Kummer

C. cookei (Bresadola) J. D. Arnold

C. tuberosa (Fries) Kummer

(ii) Glossary of technical terms

Specialised colours are placed in capitals

Adnate (of the gills or tubes), broadly attached to the stem at least for one quarter of their length. See p. 267.

Adnexed (of the gills or tubes), narrowly attached to the stem by less than one quarter of their length. See p. 267.

Amygdaliform (of the spore), almond-shaped.

Amyloid (of the spore-walls, spore-ornamentation or hyphal walls), greyish or bluish or blackish violet in solutions containing iodine.

Apiculus (of the spore), the short peg-like structure at the basal end of the spore by which it is attached to the basidium. See Fig. 5, p. 15.

Arcuate-decurrent (of the gills or tubes), curved and extending down the stem. See p. 267.

Ascus, a clavate to cylindrical or subglobose cell in which the (asco-) spores are borne, usually in eights.

Basidium, a clavate or subcylindrical cell on which the (basidio-) spores are borne, externally on stalks. See Fig. 5, p. 15.

Cap (of the fruit-body), that structure which bears the spore-bearing layers beneath it (= pileus).

Caespitose (of the fruit-body), aggregated into tufts.

CINNAMON-BROWN, the colour of cinnamon powder obtainable from the grocer.

Clavate (of the stem, or cystidia), club-shaped.

Convex (of the cap), curving outwards. See Plate 9, p. 55.

Cortex (of the cap or stem), outer layers of the tissue.

Cortina, a cobweb-like veil at first connecting the margin of the cap and stem, but at maturity often only present as remnants on the stem and/or cap-margin. See p. 267.

Cystidium, a differentiated terminal cell usually on the surface and edges of the cap, gill and stem: facial cystidia occurring on the gill-face: marginal cystidia occurring on the gill-margin. See Fig. 4, p. 15.

DATE-BROWN, the colour of packed dates.

Decurrent (of the gills and tubes), with a part attached to and descending down the stem. See p. 267.

Deliquescent (of the gills, cap or entire fruit-body), changing into a liquid at maturity.

Depauperate poorly developed.

Depressed (of the cap), having the central portion sunken, and (of the tubes) sunken about the apex of the stem. See Plate 1, p. 29.

Dentate see toothed.

Distant (of the gills), greater than their own thickness apart.

Divergent (of the gill-trama in transverse longitudinal section), with the hyphae curving downwards and outwards on both sides of a central zone as if combed. See Fig. 9A, p. 17.

Ellipsoid (of the spores), elliptic in outline in all planes.

Emarginate (of the gills), notched near the stem. See Sinuate, p. 263.

Excentric (of the cap), laterally placed on the stem.

Expallent (of the cap), becoming paler when drying.

Expanded (of the cap), opened out when mature. See Plate 10, p. 61.

Fibrillose (of the cap and stem-surfaces), almost smooth but for distinct parallel longitudinal filaments (fibrils).

Fleshy (of the fruit-body), of a rather soft consistency: readily decaying.

Floccose, with loose, cottony surface; diminutive—flocculose.

Free (of the gills and tubes), not attached to the stem. See p. 267.

Frondose trees, broad-leaved trees.

Fruit-body, the whole agaric (toadstool or mushroom, polypore, etc.), as usually understood.

Germ-pore, a differentiated apical, usually thin-walled portion of the spore. See Fig. 5, p. 15.

Gill, the structure on which the reproductive tissue is borne in agarics, resembling plates.

Globose (of the spore), round in outline in all planes.

Glutinous (of the cap or stem), provided with a sticky jelly-like coating.

Heteromerous (of the cap and stem-flesh), with discrete nests of rounded cells in a background of filamentous cells: characterises members of the Russulaceae. See Fig. 10B, p. 17.

Homoiomerous (of the cap and stem-flesh), not sharply differentiated into two types of cells, although some may be swollen: characterises agarics other than members of the Russulaceae. See Fig. 10A, p. 17.

Hygrophanous (of the cap), translucent when wet, opaque and often paler on drying.

Hymenium, the superficial layer of cells in which basidia occur. See Fig. 9A–D, p. 17.

Hyaline, appearing as if clear glass.

Hypogeous, growing under ground.

Hypha, a fungus filament composed of a chain of several cells; plural— hyphae; adjective—hyphal.

Inverse, (of the gill-trama in transverse longitudinal section), with the hyphae curving upwards and outwards on both sides of a central zone. See Fig. 9B, p. 17.

Irregular (of the gill-trama in transverse longitudinal section), lacking any clear pattern as to hyphal arrangement. See Fig. 9D, p. 17.

Mealy, covered in powdery granules, resembling meal.

Mycelium, a mass of fungus-filaments (hyphae).

Mycorrhiza, a symbiotic association of a fungus and the roots of a higher plant.

Non-amyloid (of the spore-wall, spore-ornamentation and hyphal walls), remaining uncoloured or becoming yellowish in solutions containing iodine.

OCHRACEOUS, bright clay-colour: colour of ochre (yellow-brown).

OLIVACEOUS BROWN, a dull clay-brown with an additional but distinct hint of dirty green.

Plano-convex (of the cap), regularly rounded although almost flat. See Plate 13, p. 67—adult fruit-body.

Pruinose (of the cap and stem-surfaces), finely powdered.

Pubescent (of the cap and stem-surfaces), with short, soft hairs.

Putrescent (of the fruit-body), soft and very easily decaying.

Pyriform (of the spore), pear-shaped.

Regular (of the gill-trama in transverse longitudinal section), with hyphae showing no distinct curvature and practically parallel to the gill-surfaces. See Fig. 9C, p. 17.

Remote (of the gills or tubes), separate from the stem by a zone of cap-flesh. See p. 267.

Resupinate (of the fruit-body, spore-bearing tissue facing outward and attached to support by what would have been the cap had the fungus been a normal agaric.

Ring, a girdling veil on the stem. See p. 267.

Rugulose (of a surface), covered in small wrinkles.

RUST-BROWN, the colour of rusty iron.

Saprophyte (of an organism), using dead material for active growth.

Scurfy (of the cap and stem surfaces), with small irregular loosely attached scales.

Sessile (of the fruit-bodies), lacking a stem.

Septate (of the structural units of the fruit-body), with cross-walls; septum—cross-wall.

Sinuate (of the gills), having a concave indentation of that part of the edge nearest the stem. See Plate 32, p. 111.

SNUFF-BROWN, a dull dark clay-brown said to resemble the colour of snuff.

Spore-print (or deposit), the mass of spores obtained by allowing the fruit-body to discharge its spores at maturity.

Stem (of the fruit-body), that structure which supports the cap (= stipe).

Sterile, a tissue or structure not involved in the reproductive process, or failing to take part.

Sterigma, the point-like structure at the apex of the basidium actually bearing the spores.

Striate (of a surface), having minute furrows or lines.

Subdecurrent (of the gills or the tubes), having the gill-attachment extending slightly down the stem. See p. 267.

TAWNY, sand-coloured.

Tomentose (of the cap and stem surfaces), densely matted and woolly.

Toothed (of the gills or cap-margin), as if with teeth (= dentate).

Trama (of the gills), the tissue between the layers bearing basidia (hymenia).

Umbilicate (of the cap), having a central, small depression. See p. 267.

Umbonate (of the cap), provided with a broad, flattened, raised centre (the umbo).

Uncinate (of the gills), emarginate, but with a long descending decurrent tooth because the cap does not expand. See Plate 14, p. 69.

Veil, a general term for the tissues which protect the whole or part of the developing fruit-body.

Viscid (of the cap or stem), very slippery to the touch.

Volva, a persistent cup-like structure at the base of the stem. See p. 267.

Waxy (of the gills), lustrous because they are thick and watery.

Illustrations

Text-figures and line-drawings of the greater number of the fungi mentioned in the text have been included in the book. It is impossible to supply colour pictures of a high quality in a book such as this without raising the price of the publication astronomically. The plates in six easily obtainable popular books have been used to represent whenever possible the fungus described in the text, as accurate colour illustrations are very useful in identification. The titles of these books have been abbreviated for clarity.

Abbreviations for illustrations used throughout the text

F—Findlay, W. P. K. (1967), *Wayside and Woodland Fungi*, London.
Hvass—Hvass, E. & H. (1961), *Mushrooms and Toadstools in Colour*, London.
LH—Lange, M. & Hora, F. B. (1963), *Collins Guide to Mushrooms and Toadstools*, London.
NB—Nicholson, B. E. & Brightman, F. H. (1966), *Oxford Book of Flowerless Plants*, Oxford.
WD—Wakefield, E. & Dennis, R. W. G. (1950), *Common British Fungi*, London.
Z—Zeitlmayr, L. (1968), *Wild Mushrooms*, London.

(iii) Fairy rings

Object: To assess the annual radial growth of fairy-rings and to correlate this with any obvious environmental change.
Materials: Graph and tracing papers, tape-measures, note-book, pencil and rule, small pieces of cane about four inches long and coloured dye (e.g. Eosin solution, Janus Green).
Method: Select a fairy-ring on the school cricket pitch or hockey pitch, school lawn, local golf course or park at a time when the fruit-bodies are first visible. Carefully mark the centre of the ring by driving into the soil a piece of cane until the top is only just visible. Plot this point on graph paper and relate it to any prominent feature nearby, e.g. post, tree or hedge.

Carry out weekly observations throughout the fruiting season plotting the individual fruit-bodies on tracing paper, which is trimmed so as to make a replica of the original graph-sheet. A small dab of coloured dye placed on a fruit-body will assist one in recognising fruit-bodies from previous observations. During the fruiting season observe and plot the zones of differently coloured vegetation—devise some method of describing (and measuring) these colours perhaps by comparison with a colour-chart, printed or hand prepared. Continue observations on the ring at monthly or fortnightly intervals after the disappearance of the fruit-bodies, and record subsequent changes in the vegetation for twelve months.

This project can be continued for several years and for different species of fungus. Weather conditions may be noted simultaneously with the growth observations, or obtained from a reliable source of similar information close by. In this way not only is the increase in ring size measured but the results can be considered in the light of climatic data; fungal growth appears to be dependent on favourable weather conditions.

Further experiments:

(i) Compare the effect that different species of agaric have on the same type of vegetation.

(ii) Observe selected fairy-rings for several seasons then either apply fertilisers, particularly calcium-based fertilisers to the ring-area, or mow the vegetation. Note increase in fruit-body production, if any, changes in period of fructification or incresae in rate of ring development.

(iii) Prepare transects across the fairy-ring and observe the species of flowering plants and mosses present, the differences between species in the two stimulated zones, and the colonisation of the dead zone by annuals and later perennial grasses and herbs.

(iv) To the soil from each zone apply simple soil-dilution plate-methods for the culture and isolation of soil fungi and bacteria. Compare the results with those obtained by similar methods from soil without the fairy-ring.

(iv) Development of the agaric fruit-body

In the soil or substrate the hyphae of agarics frequently grow in close contact with each other, indeed the intertwining of such hyphae to form small knots is common in many fungi. In these intertwining hyphae, those close together divide and branch, later branching again to form a heap of tissue. The fruit-body develops from, or within, this knot and at its earliest stage is usually covered by loosely branched and irregularly arranged hyphae. To the unaided eye the primordium, for this is what such a structure or early beginning is called, appears to be enveloped in a mass of pale hyphal strands, often giving the fruit-body a woolly appearance when seated on the soil, wood, herbaceous debris, etc. If more than one primordium develops in close proximity, usually all but one abort early in development, or they remain checked in formation at this stage until those closeby have matured. Some species which grow on wood are caespitose, that is clustered together, and in these cases all or many more of the primordia develop fully and simultaneously.

Often it is possible to search and find these primordia in the fields and woods, and if they are examined under the low-power of a microscope it is possible to study how the fruit-body subsequently develops from its small beginnings and the part played by the ring and volva in the development determined. Thus the origin of the veil can be located, its development followed as well as its disintegration. When the fungus is grown in pure culture on sterile dung, or soil, or wood, or simply on artificial media prepared in the laboratory the full sequence of events can be more easily followed. This is how the professional mycologist conducts his observations. By very careful studies it has been found in recent years that the development of the fruit-body, the origin of the gills, etc. can assist in the classification of the higher fungi. Thus some species have no protective tissue around the developing gills (gymnocarpic) whilst others have one or even two, simple or complex, tissues around the developing gills or pores (hemiangiocarpic). It is these tissues which give rise to the ring, volva, cortina, etc. This most exciting part of the study of the higher fungi is illustrated in the accompanying figures (Figs. 12 & 13) along with the various

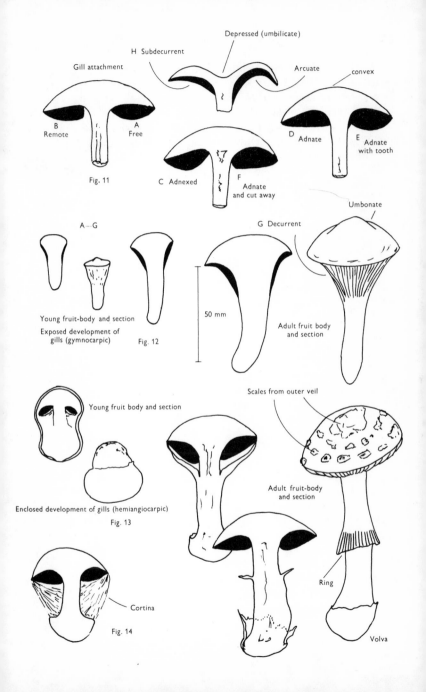

Gill attachment

H Subdecurrent

Depressed (umbilicate)

Arcuate

convex

B Remote

A Free

D Adnate

E Adnate with tooth

Fig. 11

C Adnexed

F Adnate and cut away

G Decurrent

Umbonate

A — G

50 mm

Young fruit-body and section

Exposed development of gills (gymnocarpic)

Fig. 12

Adult fruit body and section

Young fruit body and section

Scales from outer veil

Adult fruit-body and section

Enclosed development of gills (hemiangiocarpic)

Fig. 13

Cortina

Ring

Fig. 14

Volva

types of gill-attachment mentioned in the text (Fig. 11 A–H). If the agaric has two tissues surrounding it as the cap expands and matures, first the outer tissue or skin breaks leaving pieces on the stem and/or cap and then the second skin breaks as the cap expands still further. The last skin leaves remnants on the stem and sometimes bits and pieces at the cap margin. Only now can the agaric shed its spores from the fully exposed gills.

(v) References

A. Reference Texts

Some references have already been given on p. 264. Findlay, Hvass & Hvass, Lange & Hora, Nicholson & Brightman, Wakefield and Dennis and Zeitlmayr.

In addition to these the following texts are suggested:

Henderson, D. M., Orton, P. D. & Watling, R. (1969). *British Fungus Flora: Agarics and Boleti: Introduction*, H.M.S.O., Edinburgh.

Hennig, E. (1958–60). *Handbuch für Pilzfreunde*, Jena (in German).

Haas, H. (1969). *The Young Specialist looks at Fungi*, London.

Pilát, A. & Usak, O. (1951). *Mushrooms*, London.

Pilát, A. & Usak, O. (1961). *Mushrooms and other fungi*, London.

Ramsbottom, J. (1951). *Handbook of Larger fungi*, London.

Ramsbottom, J. (1953). *Mushrooms and Toadstools*, New Naturalist, London.

Romagnesi, H. (1963). *Petit Atlas des Champignons*, Bordas (in French).

Smith, A. H. (1963). *Mushroom Hunters' Field-guide*, Michigan.

Wakefield, E. M. (1954). *Observer's book of Common fungi*, London.

Watling, R. (1970). *British Fungus Flora: Agarics & Boleti, Part I*, H.M.S.O., Edinburgh.

B. General Texts

Talbot, P. M. B. (1971). *Principles of Fungal Taxonomy*, London.

Webster, J. (1970). *Introduction to Fungi*, Cambridge.

C. Journals

Bulletin Trimestriel de la Société Mycologique de France, Paris. (Official organ of the French Mycological Society.)

Coolia, Leiden. (Official organ of the Dutch Mycological Society.)

Mycologia, New York. (Official organ of the American Mycological Society.)

Schweizerische Zeitschrift für Pilzkunde. (Official organ of the Swiss Mycological Society.)

Transactions of the British Mycological Society, (Official organ of the British Society: Hon. Sec. Dr B. E. Wheeler, Imperial College of Science and Technology Field Station, Silwood Park, Sunninghill, Ascot, Berks, also publishes a Bulletin intended for the amateur.)

D. Advanced Texts

Dennis, R. W. G., Orton, P. D. & Hora, F. B. (1960). *New Check List of British Agarics and Boleti*, suppl. Trans. British Mycological Soc.

Moser, M. (1967). *in Gams Kleine Kryptogamenflora*, Band IIb Stuttgart (in German).

Kühner, R. & Romagnesi, H. (1953). *Flore Analytique des Champignons* Superiéurs de France, Paris (in French).

Rea, C. (1922). *British Basidiomycetae*, Cambridge.

Revue de Mycologie (journal) Paris (in French).

INDEX

Numbers in bold italics refer to pages with illustrations.